T0187070

Microlands

J. CRAIG VENTER AND
DAVID EWING DUNCAN

Microlands

The Future of Life on Earth

(and why it's smaller than you think)

ROBINSON

ROBINSON

First published in the United States of America in 2023 by
The Belknap Press of Harvard University Press

First published in Great Britain in 2024 by Robinson

13 5 7 9 10 8 6 4 2

Copyright © J. Craig Venter Institute, Inc, 2024

The moral right of the author has been asserted.

Important Note

A CIP catalogue record for this book
is available from the British Library.

ISBN: 978-1-47214-417-1

Printed and bound in Great Britain by Clays Ltd, Elcograf S.p.A.

Papers used by Robinson are from well-managed
forests and other responsible sources.

Robinson
An imprint of
Little, Brown Book Group
Carmelite House
50 Victoria Embankment
London EC4Y 0DZ

An Hachette UK Company
www.hachette.co.uk

www.littlebrown.co.uk

This book is dedicated, by J. Craig Venter,
to the *Sorcerer II* science and sailing teams and the
J. Craig Venter Institute scientific and logistical teams—
particularly to Karin A. Remington (1963–2021),
and to Ari Patrinos for believing in his ideas

And, by David Ewing Duncan, to his mother,
Patricia DuBose Duncan (1932–2021)

The Earth has music for those who would listen.

—Reginald Vincent Holmes, "The Magic of Sound"

Contents

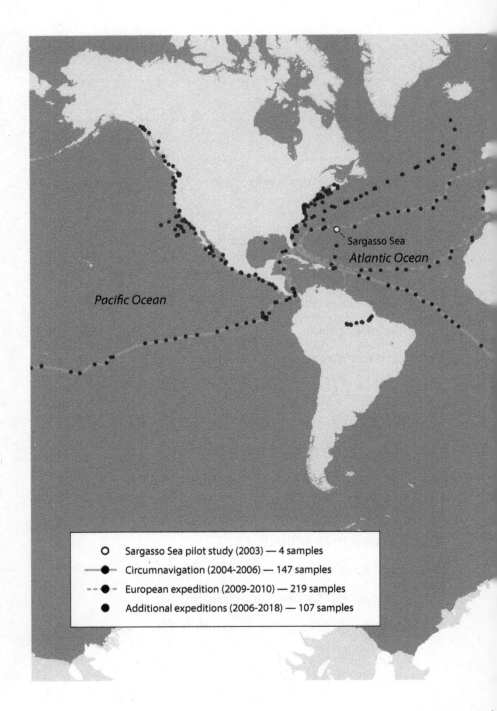

Sargasso Sea

Atlantic Ocean

Pacific Ocean

○   Sargasso Sea pilot study (2003) — 4 samples

●— Circumnavigation (2004-2006) — 147 samples

--●-- European expedition (2009-2010) — 219 samples

●   Additional expeditions (2006-2018) — 107 samples

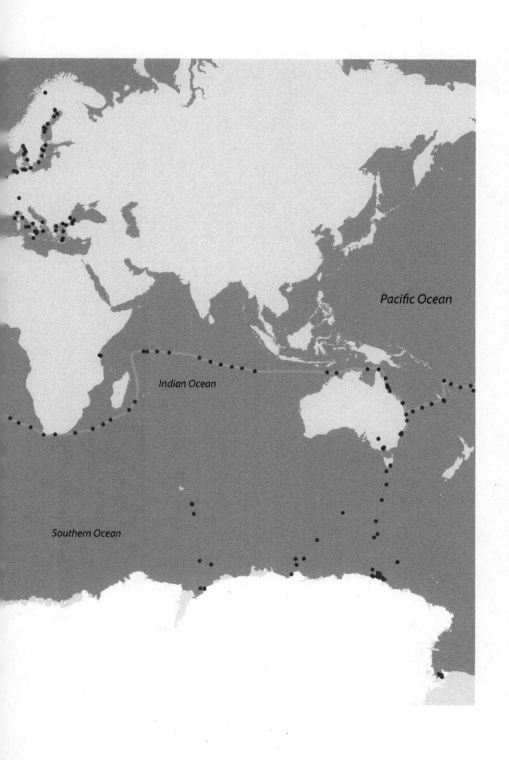

Foreword

IT WAS DURING THE 1990S THAT I FIRST MET CRAIG VENTER. Few individual encounters have had such an impact on the course of my life. As professor of virology at the Karolinska Institute and then Permanent Secretary of the Royal Swedish Academy of Sciences, I introduced Craig at many lectures in Stockholm. Each one of his talks presented impressive new advances, including the very first sequencing of a complete genome, in 1995, of the bacterium *Haemophilus influenzae,* followed by an archaean (one of the three cellular forms of life, the others being bacteria and eukaryotes)—followed, of course, by the human genome with its roughly three billion nucleotides.

The scientific community viewed with skepticism (and envy) the trailblazing efforts by this maverick of genomics, including the introduction of the "shotgun" technique for sequencing DNA. But Craig, with a qualified team of computer scientists and innovative mathematicians, continued to push the boundaries of research. He challenged the well-funded international consortium of scientists involved in the sequencing of the human genome.

Who finished first and who generated the results with highest integrity are moot points. But, importantly, without Craig's qualified and aggressive initiatives there would not have been a race to decipher the human genome, which almost certainly was finished years faster thanks to his efforts. The experiences gained by Craig and his close collaborators in the late 1990s and early 2000s changed the landscape of genomic sequencing and made possible the explosion of sequencing since, including the collection and sequencing of tens of millions of previously unknown genes described in this book.

But Craig was always looking to accomplish more than simply assembling nucleotides. He wanted to understand how life works and how things are connected. In 2009, I was kayaking with him, exploring several small rock islands along the eastern coast of Sweden. During this paddle, Craig said, "Erling, I have found the essence of life!" To which I, of course, answered: "Isn't that wonderful, tell me what it is." And the answer was: "The essence of life is naked skin against a smooth rock." To appreciate this insightful statement, one needs to know that it was the ice that covered Scandinavia until some ten thousand years ago that polished many granite rocks in the archipelago until they were smooth as silk. Craig incorporated this knowledge into an almost Zen-like realization about the beauty of life and the interaction of humans and nature. It was with this kind of awareness of the interconnectedness of the world that Craig set out to explore the microbiome of the oceans.

I was proud to participate in this venture in a very small way when Craig inquired if I was interested in sailing on *Sorcerer II* during its 2004–2005 trip around the world. I immediately said yes. As he knew, I was fascinated by the ocean and had loved to sail since I was a boy, so this was a dream come true. I had the privilege to be part of the *Sorcerer II* crew on five occasions: in the Pacific Ocean sailing from the Fiji Islands to Vanuatu; in the South Atlantic sailing from Cape Town in South Africa to Ascension Island via St. Helena; exploring the Sea of Cortez between Baja California and the Mexican mainland; sailing the North Atlantic from the Bermudas to the Azores; and touring for three days through the outermost rock islands of the Stockholm archipelago, docking at my family's summer house on Blidö. Stories from some of these sailings are presented in this book.

I'll never forget what it was like to be on board a boat the size of *Sorcerer II*—how the boat catches the heavy wind and the full sails give her a speed of ten knots or more. How, on an open sea, a pattern of interactions develops among the crew, a chamber play structured by the continuous, cyclic passing of days and nights without any contact with land. And how, in the absolute darkness of night, breaking waves become a heartbeat and the smattering of stars in a deep blue sky a silent guide.

But beyond an exciting tale of sailing across the seas, the rich story that unfolds here is one that concerns all of us as individuals and as a global community. Findings by the *Sorcerer II* expeditions have helped to reveal the great complexity of nature, particularly at the level of microbes. It is now up to us—the global community—to use this knowledge to keep the oceans and the planet healthy. And we must collect more samples. For even with the astonishing amount of information already collected, the boat's original circumnavigation represents a mere scratching of the ocean's surface.

   We must become proper guardians of the Earth for our own future, as well as for all the inhabitants that call our planet home. We still have much to learn.

Erling Norrby, MD, PhD, August 2021

Authors' Note

THIS BOOK IS A COLLABORATION between a scientist and a science writer. Primarily, it's the story of the scientist—J. Craig Venter—and his explorations into the microbiome of the Earth's oceans from 2003 to 2018. The narrative contains a multitude of thoughts and ideas from Craig and his colleagues over the past two decades on this project. It's also informed by his work in the two decades before that—most famously his seminal project in the late 1990s and early 2000s to win the race to sequence a human genome. The book also incorporates the ideas and writing style of the science writer, David Ewing Duncan—which is one reason that the book is written in the third person. Another reason is that the story includes a remarkable team and cast of characters that the scientist partnered with to make the explorations and the project's discoveries possible. These collaborations span the globe and include hundreds of researchers, funders, supporters, and others from countries and institutions and universities in dozens of countries.

In the writing, a challenge was to include Craig's distinctive first-person voice in a third-person narrative. The solution was to include his voice in quotations, in his own words, using first-person pronouns. This allowed for an account that is both *about* him as the book's primary subject and partly *by* him as a coauthor.

We use our first names—Craig and David—when we appear in the narrative, whereas last names are used to refer to everyone else, once we have provided their full names.

Microlands

Sorcerer II, under full sail, heading down the US east coast to Florida in December 2003.

PROLOGUE

Thinking Big about Small

THIS TALE STARTS BIG before it gets small.

Really small.

The big part starts like this. A man with a grizzled white beard, a deep tan, and ice-blue eyes stands alone at the helm of *Sorcerer II*, a hundred-foot-long sliver of fiberglass and Kevlar slicing through a gray-blue sea. All around him on this overcast summer day in 2018 the ocean roils in steady, meter-high combers in a modest breeze. Above the ship's bow, two huge canopies of sails swell as the wind buffets the sheets. Lulls interspersed with puffs cause the great mainsail, stretching some seven stories high, to ripple and then fill up and then go slack again, over and over, in undulating patterns that make it look almost alive.

To large specimens of terrestrial macro life like the man steering this great vessel, the sea looks empty. A liquid desert with dunes made of H_2O, a vast panorama with no beginning or end. Other macro life occasionally peeks out of the sea, organisms that we can see without the aid of a microscope. A few meters to starboard is a swarm of dolphins, cresting and arcing like charcoal-gray rainbows. They rise into the air and seem to float for impossibly long periods of time in the sky, hanging at the tops of foamy arcs as their bodies rise from the sea and gracefully slide back into the water.

The sea itself on this sunny, cloudless day off the coast of Southern Maine also seems alive. As the ship rises and falls in a steady rhythm, the ocean pitches and crests and sometimes becomes abruptly calm only to rise again into larger swells that suggest a storm is approaching but still far away.

It's hard to tell what the bearded man is thinking. Scanning the sea with a look of intense concentration, he is focused on details of wind, water, and sail, probably to exclusion of everything else, even the science he is famous for. In 1995 he was the first to sequence the complete genome of a living organism, a bacterium named *Haemophilus influenzae*. In 2000 he won the race to map the human genome, having completed the sequencing in record time using technologies and techniques that he mostly thought up and led teams to develop. In 2010 he synthesized a novel organism, creating in his lab the complete genome of a tiny bacterium made up of AGCTs of DNA not from nature but out of a bottle. When he and his team booted up this artificial genome in a cell, it came to life.

Craig Venter, the man at the helm of *Sorcerer II*, didn't perform these feats quietly or humbly. As well-known as he is for his science, Craig is also known for having a brash personality and unorthodox ideas that sometimes provoke more traditional scientists. Also a consummate risk-taker, Craig has responded to his many critics by mostly succeeding at what he sets out to do in the lab, just as he takes chances and often wins when he races sailboats, cars, and motorcycles. Sometimes he combines his passions—in this case, sailing with bio-research—and the result is a hybrid adventure like this one.

"Am I a bit of an adrenaline junky?" says Craig. "Yes."

This latest venture starts with the obvious point that the ocean is not an empty desert pulsing with water-dunes. Drop beneath the surface that separates our world of air from the liquid world below, and you immerse yourself in the abundance of macro life inhabiting these waters. Several species of whales live here at different times of the year, including humpbacks and finbacks.[1] There are numerous species of fish, too, ranging from great white sharks to Atlantic shad.[2] There are crustaceans such as those famous Maine lobsters and echinoderms like orange sea cucumbers and blood stars. Seaweeds are plentiful and include kelp, sea lettuce, rockweed, bladderwrack, and much, much more.[3] Even though overfishing and pollution have taken their severe toll along this coast, and in most of the Earth's ocean, underwater macro life is still vastly present.

Big, however, is not the point on this blustery afternoon.

This is part of an ongoing mission of discovery that, like epic scientific explorations of past centuries, spans many years. Craig launched it in 2003 when he began scouring the Earth—its land, its sky, and the oceans that cover 70 percent of the planet—in search of life so small we can't see it with our naked eyes.

It's been only three and a half centuries since Dutch lens maker and scientist Antonie van Leeuwenhoek took a microscope, then a recent invention, and created high-quality lenses that allowed him in 1676 to be the first in history to see bacteria and other organisms and particles living in a drop of water. Nearly 350 years later, with far more powerful tools available than a simple microscope, the core mission remains to explore what's out there in a micro-universe of organisms smaller than fifty microns across.* On this outing, the area of investigation is a space below the surface of the sea in the Gulf of Maine.

Earth's seas contain over 321 million cubic miles of water, according to the US National Oceanic and Atmospheric Administration (NOAA).[4] That's a volume that, translated into one-liter milk cartons, would fill over 1.3 sextillion containers—a sextillion being a one followed by twenty-one zeros. But none of these milk cartons would contain only seawater. They also would play host to an average of one billion bacteria (including the bacteria-like organisms known as archaea) and ten billion viruses

---

* A micron, or micrometer, is a millionth of a meter, or .00004 inches.

per carton, plus untold eukaryotes, algae, and fungi.[5] Every liter of water in the ocean, from surface to bottom, even when the sea floor is miles down, is swarming with such life, the microorganisms that are the true masters of our planet.

Scientists estimate that, in sheer weight, the biomass of bacteria adds up to some seventy gigatons.[6] Compare that to a mere two gigatons for all animals put together, including humans. The estimated number of all bacteria on Earth is five million trillion trillion—that's 5,000,000,000,000,000,000,000,000,000,000 (thirty zeros).[7] By comparison, there are eight billion (8,000,000,000) of us, with a mere nine zeros.

When Craig started his around-the-world ocean quest in the early 2000s—officially called the Global Ocean Sampling (GOS) expedition—microbiologists and other scientists dismissed his project to survey large swaths of the world's oceans for microbes as pointless. They were certain that there was only limited micro life in the sea. They also called it an adventure-travel vacation posing as science by a man who loves to sail. This wasn't the careful, contained, quiet work that tends to be the norm in science, they said. Rather it was the sort of wanderlust that drove explorer-scientists of the nineteenth century, as when the young Charles Darwin sailed the *HMS Beagle* mostly searching for new (macro) life. It is a style of discovery that seems random and uncontrolled to most modern researchers—but not to a risk-taker like the man with the ice-blue eyes.

At some point today, the crew will ease the sails on this luxury-yacht-turned-research-vessel as they stop and take on board two hundred liters of seawater. They will deploy instruments to measure the water's temperature, salinity, dissolved oxygen levels, and more. To collect samples the seawater will be sucked up using a pump that's dropped into the ocean and dragged beside the boat using a long pole. Once samples are on board, scientists will force the water through a series of micropore filters housed in round holders made of stainless steel, mounted on a rack in the aft cockpit of the boat.

The filters will catch microorganisms of different sizes, with only the smallest making it through to the last filter. The microbes collected on the filters will then be frozen and sent to the J. Craig Venter Institute in La Jolla, California. There, researchers will work to identify the microbes using advanced sequencing technologies, mathematics, and artificial intelligence programs that Craig and his team have helped to invent and refine over the years.

Most of these tiny organisms are tiny circular cells which, when viewed through a microscope, show no clearly distinguishing features. Others are shaped like stars, ovals, rods, and helixes. Some have hair-like projections called pili, others have shell-like coverings, spiky exteriors, or whip-like tails called flagella.

Before scientists like Leeuwenhoek began to peer through their microscopes, no one imagined that such a world existed. And although microbiology has become a major scientific discipline since Leeuwenhoek's time, much of this world has yet to be explored. Its organisms remain hidden, secret, and ignored despite inhabiting virtually every nook and cranny of our planet. As you read this, some thirty-eight trillion bacteria reside in your body.[8] They are everywhere, from the tip of your incisors to your small intestine to the alveoli that absorb oxygen in your lungs. And while most people still think of bacteria as odious pests best snuffed out with antibiotics—an unfortunate residue of the germ theory that in the nineteenth century revolutionized the identification and treatment of bacterial infectious diseases—the truth is that without them you would die sooner.

Most bacteria inside you and living all over our planet are beneficial, some even vital. For humans, bacteria help us digest food and modulate our immune system. Bacteria, it turns out, connect all living species, linking us with the soil, waterways, and atmosphere of our planet in a vast web that lives and breathes almost like a colossal organism itself—five million trillion trillion individual cells that undergird and support the macro flora and fauna that make up the world we can see, from oak trees and hummingbirds to ladybugs, Labradors, playful dolphins, and you.

Microbes, primarily bacteria, do this by secreting vital chemicals and, in some cases, by performing tasks like converting sunlight to energy and oxygen. They are agents of rot and decay as they devour and break down everything that dies, and also agents of rebirth as they recycle the raw materials of life that come from carcasses of flies, dandelions, amoebas, and humans. "Microbes process everything," says Jack Gilbert, a professor and microbiologist at the University of California, San Diego, and the Scripps Institution of Oceanography. "Without them we'd be knee-deep in our own excess excretions, and up to our ears in orange peels and pine bark. Everything that is waste or has died has to be recycled into the primary chemicals of life— carbon, nitrogen, and so forth—and microbes do most of this recycling."

As scientists discover more about these cellular factories and their genes, they are also finding potential new sources of bioenergy, pharmaceuticals, and cleaner and safer industrial chemicals. Bacteria, either naturally occurring or bioengineered, are key to producing products that include antibiotics, vitamins, enzymes, solvents, beverages, foods, and more. Scientists also use natural and bio-manipulated versions of bacteria to create alternative energy sources like hydrogen fuel and ethanol made from cellulose. Photosynthetic bacteria in the ocean are key to arresting climate change because, like trees, they absorb carbon dioxide and release oxygen into the atmosphere.

In a little-known twist to the "inconvenient truth" narrative written by former US vice president Al Gore—about how human activity has increased carbon buildup in the atmosphere—the flood of carbon and other pollutants into the environment also alters the balance of microbial species on planet Earth. Excessive carbon threatens to disrupt the ocean system supporting the phytoplankton that absorb carbon dioxide and produce as much as 80 percent of Earth's oxygen. More carbon also means increasing numbers of the microbes that live in so-called dead zones: stretches of oxygen-depleted water often drenched in nitrogen, potassium, and phosphorus from fertilizers that wash off crops and lawns. In the Gulf of Mexico, for example, a dead zone stretches six thousand square miles south of the mouth of the Mississippi River. Another one, in the Gulf of Oman at the entrance of the Persian Gulf, is ten times larger, at 63,700 square miles.[9] At their worst, dead zones don't support fish and other oxygen-breathing macro organisms at all; short of this, they cause abnormalities such as slowing or stopping the growth of shrimp.[10]

Human activity is thus undermining the work of those five million trillion trillion single cells that help to keep healthy the global ecosystem supporting life as we know it. If this sort of disruption continues, the microbes will adapt and survive as they have throughout the three-and-a-half billion years they have existed on Earth—including the early years when the atmosphere contained much more carbon than it does today. But it's highly unlikely that humans will undergo such quick adaptation.

When the scientific voyages of *Sorcerer II* commenced in 2003, microbiologists had cultured fewer than two percent of the bacteria thought at the time to exist in nature, including those that live in the oceans and waterways of Earth.[11] Before shotgun sequencing allowed scientists to identify bacterial species using their DNA, culturing was used by microbiologists to feed and grow a bacterium in a petri dish so that they had enough replicated cells to identify it.

Understanding better and discovering more of that 98-plus percent of bacteria never cultured was the objective of the scientists aboard *Sorcerer II* for its various voyages from 2003 through 2018. It was a purpose made possible by Craig's outlandish idea to use genetic sequencing on a global scale—to go big in exploring the world of the very small. Along the way, they gathered microbes from locales as far-flung as the Galapagos Islands, the Panama Canal, Tasmania, Glacier Bay National Park, the Baltic Sea, and the Sea of Cortez—and also, off-ship, from fetid ponds, Antarctica, deep mines, the Amazon, volcanic vents, and the atmosphere above New York City.

The expeditions continued until 2018, with the total catch in genes discovered numbering well over a hundred million. (Compare this to humans, who have around twenty thousand genes). Craig and his colleagues have deposited billions of base pairs—the pairings of adenine (A) with thymine (T) and cytosine (C) with guanine (G) that are bonded together like rungs on a ladder to make up the double-helix strands of DNA—in public databases like GenBank, maintained by the National Center for Biotechnology Information, and CAMERA, funded by the Gordon Moore Foundation.

Yet merely collecting x number of samples, microbes, genes, and base pairs was never the point. The global sampling expeditions were part of the journey that Craig began when he sequenced the first cellular genome, that of *Haemophilus influenzae*, in 1995. That's when he began using bacteria as an experimental organism to refine shotgun sequencing and other technologies and processes, while also vastly accelerating progress toward identifying and understanding the structure and functions of genomes and individual genes in multiple species, including humans. This would in turn lead to his seminal experiments in building synthetic organisms in the lab, seeking to understand the function of bacteria and what constituted life at the DNA level.[12]

On this gray afternoon in the Gulf of Maine in 2018, the fifteen-year *Sorcerer II* expedition is nearing its end as Craig gazes once more at the vast sea churning and swelling around him. "I have to force myself to imagine that every milliliter of sea out there has a million bacteria and ten million viruses," he says. "I still see it as this wondrous, beautiful thing, but what it really is, is a massive, living soup—one that we are still exploring, as we try to learn the great secrets it holds about life on Earth."

PART I

IN SEARCH OF MICROBES

1

Sargasso Sea Surprise

*A sudden, bold, and unexpected question doth
many times surprise a man, and lay him open.*
—FRANCIS BACON

ON MAY 13, 2003, *Sorcerer II* was slipping elegantly through the blue-gray waters of the Sargasso Sea, twenty-six miles southeast of Bermuda. The sky was clear and the sea calm on a day that everyone on board remembers as being hot and muggy. In the back of the vessel, twenty-nine-year-old Jeff Hoffman—a tall,

broad-shouldered microbiologist with sun-bleached, buzz-cut
hair, and a champion swimmer from Louisiana—was preparing to take the first ocean microbe samples ever procured
from the stern of *Sorcerer II*. With a small team of scientists he
was setting up a pump and instruments that would be deployed into the sea at or very near the longitude and latitude
coordinates of 32°10′N 64°30′W. This was the location of Hydrostation "S," a deep-ocean mooring where scientists had been
taking regular oceanographic samples since 1954. Every two
weeks, submersible bottle kits had been dropped from research
vessels to collect water at various levels below the surface.[1] Instruments also collected data on temperature, salinity, and
dissolved oxygen levels.[2]

Hydrostation "S" sits near the western edge of a vast sea
that runs from Bermuda in the west toward the coast of Africa,
stretching about halfway across the Atlantic. Bordered by strong
and steady currents that surround it on all sides like the liquid
walls of a watery fortress—the Gulf Stream to the west, the North
Atlantic Current to the north, the Canary Current to the east,
and the North Equatorial Current to the south—the Sargasso
has sometimes been called an "ocean desert," because these
currents block the flow of ocean nutrients that support abundant macro life in other parts of the Atlantic. The Sargasso Sea
is also notable—and was named—for the abundance of seaweed
of the genus *Sargassum* floating on its surface. It's a seemingly
infinite, saturated, mottled mat that looks like piles of dark,
green-brown mush.[3]

As of 2003, when *Sorcerer II* set out on that steamy summer
day, it was still assumed by most scientists that low-nutrient
concentrations in the Sargasso Sea meant that microbes would

be scarce, too. Before this expedition, ocean microbiologists had identified precious few of the tiny organisms living there. But was the Sargasso really a life zone for so few microbes, or had scientists missed some or most of the species living there?

To answer this question, the scientists on board *Sorcerer II* would be sending back an unprecedented trove of microbes to a lab in Maryland to be genetically sequenced using shotgun genomic sequencing—a relatively new process that Craig had first used in the 1990s, including at Celera, the company he co-founded in 1998. Shotgun sequencing works by extracting the DNA from an organism (or from a dozen or hundreds or thousands of organisms) and literally blasting it into short, random fragments. DNA-sequencing machines then capture each fragment, roughly five hundred base pairs long, and spell out its genetic code in terms of the A, G, C, and T bases that are the four building blocks of DNA. Finally, computers find the overlaps in these fragments—stretches of DNA that match by at least sixteen base pairs—and digitally assemble them into genes and genomes, which are then compared to known DNA sequences.

Previously, geneticists had used a more time-consuming and expensive process of painstakingly slicing up long stretches of DNA into shorter ones for sequential analysis. It had been in the 1990s that Craig pioneered the use of shotgun sequencing to decode bacterial genomes, starting with *Haemophilus influenzae*—and this process of analyzing random fragments had since been adopted universally to sequence single-species organisms, from bacteria to humans, mice, monkeys, and more. No one, however, had yet tried using the technique on a sample known to contain many different microbial species, possibly numbering in the thousands or more. Nor did anyone—including Craig—have any idea that over the next fifteen years the *Sorcerer II* expeditions, using shotgun sequencing, would discover millions of novel genes, many times more than the twenty thousand or so in the average human genome.

Taking a quick glance around at the flat, blue sea that stretched off in every direction that hot afternoon on the Sargasso Sea, Hoffman wiped the sheen of sweat from his forehead and pondered again why he was there on the deck of *Sorcerer II*. A few months earlier, he had been studying for a PhD in microbiology at Louisiana State University in Baton Rouge, Louisiana. Not the microbiology of the sea, but of the desert. "I was used to sifting sand and soil, not water," Hoffman later recalled. "And yet there I was taking samples in the middle of the ocean. Was I nervous that I would do this right? Hell yes."

Captaining the hundred-foot yacht that day was Charlie Howard, a ruggedly handsome former electrical engineer with unruly dark-brown hair. Standing barefoot in baggy shorts at the helm of *Sorcerer II*, fifty-four-year-old Howard watched the preparations behind the ship's wheel as he began the process of slowing down the great vessel. With the ship turning directly into the light wind, causing the sails to slacken and flap, he flipped several switches to activate the hydraulic winches. Purring, these machines slowly spooled the mainsail into a compartment in the boom and other sails around the forestays supporting the mast at the bow of the boat. The skipper then turned on the Cummins engine and used bow and stern thrusters to maneuver the ship into the exact location where the scientists wanted to collect their samples.

Howard had been the captain of *Sorcerer II* since shortly after the ship arrived nearly new from New Zealand in late 2000 and he remained the skipper until the expeditions ended in 2018. Originally from Canada, he had a degree in electrical engineering, but had fallen in love with sailing. Just before *Sorcerer II,* he had been captaining boats in the Mediterranean, working as an engineer for yacht owners that he referred to, with understated humor about their scant sailing prowess, as "gentlemen enthusiasts."

"I had two friends email me telling me about this scientist guy who had a hundred-foot sailboat and was looking for a captain," recalled Howard. "I called him from Palma Mallorca and had a good conversation, and then when I arrived back in the States a week later I went by his house on the Cape and met with him and his wife to discuss *Sorcerer* and sailing in general. Craig was a guy who very much loves sailing and living life, and his enthusiasm and passion for sailing made it an easy decision."

Howard was hired both for his acumen around boats and his training as an engineer. The latter would prove especially critical when the *Sorcerer II's* engine broke down and briefly caught fire, stranding the vessel thousands of miles from land in the South Pacific with no wind, low fuel, and supplies that would not last forever. "Charlie as captain was also a scientific collaborator," recalled Karla Heidelberg, one of the researchers working and sweating with Hoffman during the first sample taken in the Sargasso Sea. "He helped us position the boat and made sure the equipment was working."

A microbiologist and oceanographer with a PhD from the University of Maryland, Heidelberg is a petite woman with wavy, dark hair and an even-keeled energy. On that May morning she was just finishing a stint as a scientist for the US State Department. This was useful experience for a project that would

venture into the territorial waters of numerous countries, including Bermuda, in an era when scooping up biological samples was hardly a trivial matter. Her familiarity with treaties
and complex protocols was especially valuable in dealings with
nations that were leery of, if not borderline hostile to, American
scientists gathering species that might hold the keys to novel
industrial chemicals, drugs, or energy production and storage
processes. As the treaties spelled out, countries—particularly
those in the developing world that had seen their natural resources exploited by colonialists and other foreigners in the
past—wanted to be sure they got compensated for anything
valuable discovered on the land or in the sea within their borders and territorial waters.

Karla's husband was also part of the crew. Bearded with
round glasses, John Heidelberg was a senior marine biologist at
Craig's institute, then based in Maryland.* Also on board was
Tony Knap, a British oceanographer who then headed up the
hundred-year-old Bermuda Institute of Ocean Sciences (BIOS),
originally called the Bermuda Biological Station for Research.
In a twenty-five-year stint as director, Knap had built up BIOS
from a small, isolated biological station to a major oceanographic research center that attracted top scientists and had
generated thousands of peer-reviewed journal articles. BIOS

---

* By 2003, Craig had formed separate organizations focused on
several different purposes, including the Institute for Genomic
Research, founded in 1992 when he left the NIH. All of these were
later consolidated into the J. Craig Venter Institute (JCVI). To avoid
confusion, in this book we will generally refer to them not by their
original names but simply as JCVI.

was a cosponsor of the *Sorcerer II* Sargasso Sea expedition and Knap was a key collaborator in the hunt for ocean microbes.

Another key character in the Sargasso Sea voyage was *Sorcerer II* itself. Just a day before the first sampling in the Sargasso Sea, the ship had arrived from the Caribbean island of St. Barts, having sailed the thousand miles to St. George's Harbor in two and a half days. According to Hoffman, who was already in Bermuda, *Sorcerer II* arrived in late afternoon on May 12, 2003. "We immediately went to a bar," said Hoffman. "I think it was called the Hog Penny. We partied late, which often happened when we made port. Honestly, I was probably a little hungover on that first day of sampling."

Sorcerer II was designed by the famed Argentinian yacht designer Germán Frers and built in New Zealand by Cookson Boats. Frers's team of engineers had designed fast, sleek boats for decades, including winners of many major races around the world. In 1997, Frers famously designed what was then the largest sloop in the world, the 156-foot *Hyperion,* for tech mogul and Netscape founder Jim Clark. Using the Frers design, Cookson built *Sorcerer II* in 1998, constructing it to be not only beautiful but tough. Made of fiberglass, Kevlar, and light but strong core materials, and coated with advanced chemical laminates, *Sorcerer II* was designed to handle latitudinal weather extremes from equatorial heat and humidity to the frigid waters of the northern seas—almost everything short of the serious ice of the Arctic and Antarctica.

On the wide, gently convex deck of *Sorcerer II,* the bow is dominated by stainless steel rod rigging supporting the mast and by an inflatable boat—a Zodiac—that can be dropped into the water using lines from the top of the mast and a hydraulic winch. Behind this, the large, main cockpit is partly covered by a sturdy dodger that can stand up to some nasty weather. This frame-supported canvas sits just aft of the middle of the ship, looking elegantly and artistically off-center when viewed from a distance. At the aft end of the center cockpit, with its cushioned seats and broad teak table, is the helm station: a pedestal with a large steering wheel surrounded by banks of switches and instruments. In the stern is another, deeper cockpit. This is where Jeff Hoffman was working, and where filters, hoses, and various instruments for microbe collection had been installed.

Below decks was a beautiful and comfortable main cabin with overstuffed chairs, a long table with bench-style seats and chairs bolted to the floor for when the seas got rough, and a second helm station with a large monitor that displayed sonar readings and the GPS position of *Sorcerer II*. (It was here that a crewperson or guest assigned to a four-hour night watch typically sat when it was too cold or inclement outside to monitor the helm in the outer cockpit.) In the bow below deck were two crew cabins, each with two bunks; just aft of the crew quarters was a well-stocked galley with stovetop and oven, full-sized refrigerator and freezer, and small dining table. Off the galley was an office or lounge where scientists could work at computers or use the large microscope that for several years was kept on board.

Sorcerer II was not the first craft Craig had owned since learning to sail as a young man. "I had a series of smaller boats as I learned to ocean sail and work my way through college," said Craig. "My first ocean voyage was in a Cape Dory 33 named *Sirius*, after the Dog Star, from Annapolis to Bermuda. The boat before *Sorcerer II* was also a Frers design, called just *Sorcerer*. It was eighty-two feet long and purchased from Gary Comer, the founder of Land's End and a former Olympic sailor."

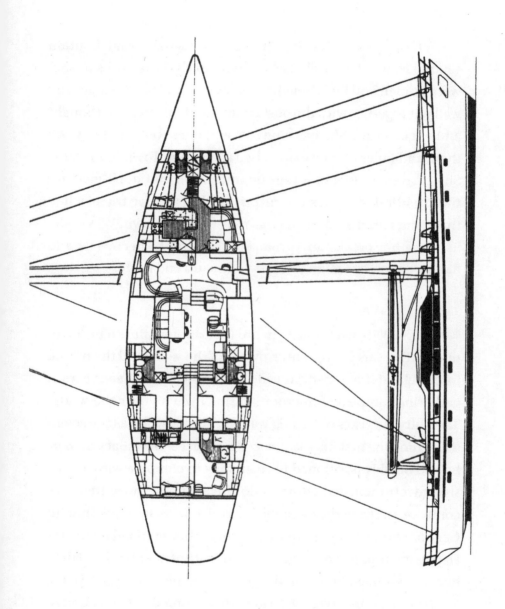

The blueprint of *Sorcerer II*, a ninety-five-foot performance cruiser built by TP Cookson in New Zealand in 1998.

"*Sorcerer* is not too big and not too small," said Captain Charlie Howard, describing his vessel, like sailors do, as if "she" were human and had thoughts and emotions. "She is smart and well put together with the best components and a lot of thought and engineering. She has long legs and once took us almost six thousand miles on one load of fuel from Cape Town to Antigua. She thrives on lots of attention and when you don't give her the attention, she gives you surprises. She is a good friend when the going is rough, and she has never let me down. When you listen to her creaks and groans and attend to her needs and maintenance, she gives you great pleasure and pride."

⋎

ONE FINAL CHARACTER who was on board that day remains to be introduced. This was Craig Venter, owner of the yacht and the primal force behind this expedition and the subsequent fifteen years of gathering samples of microorganisms from all over the world. Then fifty-six years old, Craig was world-renowned as the scientist behind the first fully sequenced human genome. Back in 1998, he had brashly positioned his company to compete with a massive government-led effort to do the same—a move that was controversial from the moment Celera was launched. The Human Genome Project, an international consortium funded by the US Department of Energy and National Institutes of Health (NIH), had already been pursuing the goal at that point for eight years.

The rivalry and even bitterness that erupted between Celera and the public-sector project at times became front-page news in the late nineties as they competed to be first to finish. The race ended officially in June 2000 when President Bill Clinton in a White House ceremony declared a tie. "It wasn't really a tie—

we won," insisted Craig. "But we went along with the tie pro-nouncement to make everyone happy." In the same spirit of joint celebration, both teams published their versions of the nearly complete human genome on the same day, February 16, 2001. Celera's version, with Craig as lead author, appeared in the journal *Science*.[4] The International Human Genome Sequencing Consortium published its version in *Nature*.[5]

Celera's sequencing of the human genome cost around $100 million and required only nine months to complete. The public effort took more than a decade and, between US and European funding, cost nearly $5.5 billion (in 2022 dollars). In 2004, when Craig was interviewed on board *Sorcerer II* for a Discovery Channel special, he was asked about the risk he and the Celera scientists had taken in that earlier high-profile competition. "If our experiment had failed," Craig told the film crew, "it would have been one of the most spectacular burnouts in the history of science. The whole world was watching everything we did then, and the risk of failure was enormous. But it quite honestly never occurred to me—I was just certain it would work. It's only when I look back now that I see literally thousands of reasons why it should have and could have failed."[6]

"I think the one thing that is pertinent for my career is not being afraid to take risks," Craig continued. "We all see it every day in people who wasted their lives because they want to take the safe route. You know, I feel the same need for security that everybody has, but I force myself to overcome this to go out and try to do new things and take risks."

By venturing out on a yacht-turned-research-vessel, Craig was making another huge bet and believing his high-tech and high-risk version of science would pay off—this time by proving that microbes were far more diverse and abundant in the oceans than anyone had previously guessed. He also had a hunch that, by discovering and identifying a slew of novel microbes, this adventure would not only transform human knowledge about the oceans but also provide fresh insight and ideas for industry and energy production. It was a gamble that few in the field thought would amount to anything significant. Back then, to the extent that microbiologists were doing such analysis, they focused only on small-scale samples taken from an estuary or cove, or on one location on a map, such as Hydrostation "S." Some had talked about larger-scale efforts, but none before the *Sorcerer II* expeditions had the moxie or the resources to attempt.

In May 2003, Craig had both. His funds were in part the proceeds from Celera and an earlier company he cofounded, Human Genome Sciences. Stock and cash from the two companies had been donated to The Institute for Genomic Research (TIGR), his first institute in Rockville, which was, along with other institutes that Craig founded, eventually folded into the J. Craig Venter Institute (JCVI). Proceeds from stock sales also paid for *Sorcerer II.*

Now, as Hoffman and the others were busy in the stern, Craig stood next to Charlie Howard on *Sorcerer II's* upper deck watching the captain maneuver the big boat into position. Later, Howard would recall how he and Craig operated onboard the vessel: "I would describe it as Craig was the admiral and I was the skipper. Craig had the broad vision of where he wanted to go and what he wanted to do, and I implemented it on a day-to-day level with planning, preparation, and execution." He remembered that "we got along well because we both respected each other, and we were complementary. Craig loved to be on the boat and doing the sailing, but still had other commitments he had to attend to at home and so would have to come and go."

Beardless, fit, and tan in a loose, sleeveless muscle shirt, Craig took in all the activity on his vessel, his ice-blue eyes as piercing as ever. "With Craig, everything was go, go, all the time," said Jeff Hoffman, recalling the early 2000s. "He was very intense, which of course made me want to do a great job." Charlie Howard agreed: "What was Craig like then? Way less laid back, with lots of balls in the air and things going on, which he loved."

Many who knew Craig during this period also describe him as unflappable, especially in the face of criticism. A year later, science writer James Shreeve, author of *The Genome War*, would write a cover story about Craig in *Wired* that portrayed him as both driven and calm. Traveling to visit Craig and *Sorcerer II* in the South Pacific during the 2004–2006 global circumnavigation, Shreeve described Craig as "bald, bearded and buck naked." The beard was new since he'd graced the covers of *Time* and *BusinessWeek* during the Celera era. "It makes him look younger and more relaxed—not that I ever saw him looking very tense, even when the genome race got ugly and his enemies were closing in."[7] Heather Kowalski, who was Craig's publicist at Celera and later became his wife, remembered differently. According to her, Craig kept up an outwardly calm demeanor during the Celera days but was frequently under intense stress.

Certainly, by the time *Sorcerer II* was scooping up its first sample in the Sargasso Sea, Craig had been through a dramatic year of ups and downs. Besides the recent publication detailing the human genome in *Science,* he was still recovering from getting fired by Celera's board in early 2002. Craig hadn't been getting along with Tony White, the CEO of Applera, Celera's parent company, because White wanted to turn Celera into a pharmaceutical company to justify its high stock valuation. Craig's vision was to provide pharmaceutical companies and others with analyzed and annotated genomic information—to be the "Bloomberg of biology," as Kowalski put it. In the end, neither approach saved Celera from the stock market crash in 2001 and the realization among geneticists and in industry that sequencing the human genome was just the first step on a long road to developing and selling novel therapeutics. Much discovery was still required of what all those genes did.

Celera under Craig had gone through an extraordinary boom cycle since its founding in 1998. Propelled by the hype around the genome project—and the overheated, dot-com-fueled markets of the late 1990s—the company had skyrocketed to a $24 billion valuation only to plummet to less than a billion by 2002, the year Craig left.

"Getting fired from Celera was a real shock," recalled Craig years later, even though he had told a board member that he wanted to go back to his research institute. "The intensity level was probably equivalent to running a presidential campaign or something similar. People made billions off Celera. But there was an equal number that lost billions. I was the first biotech billionaire on paper. And I joked, 'I made a million dollars the hard way. I made a billion fast—and lost it just as fast.'"

Craig admits, however, that he was never all that interested in the business side of the company. "I was in it to make scientific history by sequencing the first human genome," he said, "which is why I gave my stock to the institute." The $150 million brought in by those shares funded TIGR and then JCVI, and continues to fund the institute today.

In a way, the May 2003 sampling expedition off the coast of Bermuda signaled the end of a period that had begun with Craig's departure from Celera when he stayed largely out of public view, spending most of his time sailing around in the Caribbean. He says he needed time to recharge. "He needed to regroup after Celera," said Juan Enriquez, an investor and friend based in Boston who visited the *Sorcerer II* expeditions many times over the years, and would later cofound a company with Craig, Synthetic Genomics (since renamed Viridos). "When he has setbacks, he focuses on stuff he can do and manage, like sailing. Somewhere in the back of his head, this idea was also happening about finding microbes in the ocean."

⌄

THE DECISION TO DEPLOY *Sorcerer II* to collect samples in the Sargasso Sea, and ultimately all over the world, was finalized in late 2002 in another bar in Bermuda, called the White Horse. A whitewashed, British-colonial-looking structure flush with the harbor in St. George's, the White Horse is a cozy pub with exposed beams, lots of dark wood, and a menu heavy on seafood, shepherd's pie, and British ales. Craig, however, preferred scotch, and several were consumed that afternoon in 2002 as he chatted about microbes and big ideas with Tony Knap, Jeff Hoffman, and the Heidelbergs.

The real impetus behind the Sargasso expedition, however, and Craig's obsession with microbes, was years earlier and

far away from the White Horse Pub in Bermuda—during the
Vietnam War, when Craig was a Navy corpsman treating infectious diseases.

In 1965, before being deployed to Da Nang, Craig served in
the infectious disease unit of the Naval Medical Center in San
Diego. This is where soldiers returning from Vietnam came to be
treated for malaria, tuberculosis, cholera, hepatitis, meningitis,
and fungal infections. The training profoundly affected Craig.
"Seeing these diseases ended up influencing every stage of the
rest of my life," he later reflected, and the experience "was definitely on my mind later as I was thinking about the *Sorcerer*
project." Starting in the mid-1990s, he would go on to sequence
every disease microbe that he had first seen in these patients.

This included the bacterium that causes Meningitis B, which
he first saw in the Navy and years later sequenced in collaboration with the famed Italian molecular biologist Rino Rappuoli—
an effort that paved the way to the development of the first
Meningitis B vaccine.[8] "It became immediately clear that this
was the new way for making vaccines," said Rappuoli about his
work with Craig, "which I call reverse vaccinology, because
we start with genes. This was the first time you didn't need the
pathogen, and could go backwards from the information in the
genome."[9]

One other major through-line leading to the 2002 discussion
at the White Horse Pub was Craig's relentless quest to sequence DNA faster and cheaper. "The *Sorcerer* project scientifically started in '95," said Craig, when his team used the shotgun
method on *Haemophilus influenzae*, the first organism ever to
be fully sequenced. "Actually, we did two," he said, "*Haemophilus
influenzae* and *Mycoplasma genitalium*"—projects described in

more detail in Chapter 2. The purpose of this early effort was first to prove that shotgun sequencing worked, and then to see what the method revealed about the inner workings of these cells. "We didn't set out to sequence the first genome in history," he said later. "We set out to try an experiment using whole-genome shotgun sequencing, which we believed would speed up full-genome projects from taking decades to down to less than a year."

Craig also wanted to peek into and explore a world that no one had ever seen—all the DNA of a single organism. "Doing the first two bacteria genomes," he said, "what we discovered as we studied them and compared them is that they are much more diverse and complicated than people thought they would be." The sequencing triumph was very rewarding, he said, "but at the same time, it was a point of frustration because I could tell by looking at the first complete map of all the genes associated with these microbes that they were trying to tell us far more than we could understand."

This made him wonder what scientists would find if they were able to collect and test, using shotgun sequencing, large numbers of microbes from humans, the air, the soil, lakes, rivers, and oceans. "It's only by having all the genetic code of most or many of the other species on this planet that we can truly begin to understand how life came to be what it is today."

During this same period, Craig was reading studies about microbes in the soil, air, and ocean that suggested far less diversity and complexity than he believed existed in the environment. Most irksome to him were studies claiming that microbes in the Earth's oceans numbered fewer and were less diverse than microbes in the soil. "This just didn't ring true to me," he said.

"For one thing, ocean algae produce on the order of 40 percent of the Earth's oxygen. And microbes in the soil had been studied a lot more."

Craig's skepticism also derived from what were then the prevailing methods for identifying bacteria. The oldest approach, first used in the early twentieth century, was to take a sample of bacteria, try growing it in quantity in petri dishes by feeding it a protein-rich broth of "growth medium," and then inspect the batch of microbes under a microscope and attempt to ID them. Unfortunately, since some 98 percent of bacteria fail to grow in medium, very few could be replicated in the quantities required to observe and identify them. As for the rest, with no way to grow them in the lab, it was as if they did not exist.

In the 1970s, researchers began to use genetics to identify bacteria, which was a game changer. Sequencing remained expensive and the methods painstakingly slow, yet progress was made in 1977 when the famed microbiologist Carl Woese, of the University of Illinois Urbana-Champaign, discovered how to identify many bacterial species by sequencing tiny stretches of DNA from a gene called 16S rRNA.

This gene exists in every bacterial species on Earth and has characteristics that are highly conserved in evolution—meaning it hasn't changed much in billions of years, except for minute differences in coding that vary from species to species. Acting as a kind of genetic fingerprint or barcode, 16S rRNA can tell scientists if a microbe is one species or another, or a species not previously seen, without their having to sequence the rest of the organism's genome.

The use of 16S rRNA sequencing to characterize species was a critical discovery at the time not only because it allowed far more microbes to be identified than the old broth-in-a-petri-dish method, but also because bacteria are notoriously difficult at times to tell apart given how fast they mutate. (They tend to divide every few hours, meaning they can adapt and mutate quickly.) They also sometimes borrow and share genes laterally—that is, not through reproduction. This slipperiness of genes and mutations in microbes can make it difficult to tell species apart even with 16S rRNA identification, a problem we'll discuss more in Chapter 2.

Craig's issue with 16S rRNA wasn't just this gene slipperiness, however. For him, what had rankled for years was that 16S rRNA doesn't tell us anything about what a bacterium's genes do. It's like knowing the names of, say, ten different people without knowing anything else about them. "The problem was that 16S rRNA didn't account for everything," said Craig, "and it didn't tell you anything about the function or nature of the organism. That's what I was interested in."

This is where shotgun sequencing came in, a technique that sequences an organism's entire genome, not just a barcode of DNA. The shotgun method developed by Craig in the 1990s—and used throughout the *Sorcerer II* project—also deploys sophisticated computer algorithms to assemble and analyze genomes, offering up complete genetic details, not just single genes.

ᐁ

AFTER HE LEFT Celera in 2002 and took his break to get his head back together, Craig emerged from sailing around in the Caribbean ready to return to microbes. By collecting large numbers of them to sequence and to study, he was confident he could prove that microbial life forms were much more numerous and diverse than the scientific community believed. He also wanted to definitively demonstrate that shotgun sequencing was a powerful method to use not just on single organisms but on mixtures of organisms—even thousands of millions of them in, say, two hundred liters of Sargasso seawater, collected, let's say, from a hundred-foot yacht.

"As word got out about what Craig was proposing back then," said Andy Allen, a microbiologist and oceanographer who now has joint appointments at the Scripps Institution of Oceanography and JCVI, "people in the oceanographic community were horrified. They thought, here was this rich and famous guy who wanted to hang out on his sailboat and collect a few samples of microbes. Who was he to be telling scientists who had worked with ocean microbes for years and decades what to do?"

"There were people who were devoting their entire careers to finding a few bacteria and trying to understand them," said Craig about the naysayers. "They thought I was nuts by thinking that we should set out to take as many samples as possible, having no idea what's in there, and then use shotgun sequencing to find out what we've got."

Craig did have his early supporters. These included famed Harvard biologist and bestselling author E. O. Wilson, who served on the scientific advisory board of the *Sorcerer II* expeditions. "We're talking about an unknown world of enormous importance," Wilson told *Wired* reporter James Shreeve.[10] "Venter is one of the first to get serious about exploring that world in its totality. This is a guy who thinks big and acts accordingly."

Another longtime supporter was Ari Patrinos, the former director of the US Department of Energy's Biological and Environmental Research Program, a major funder of Craig's projects over the years. "Craig is a very mercurial and a very tough personality" said Patrinos, "which is not a negative trait, as far as I'm concerned. I think it's always been a tremendous strength of personality and commitment to the ideas that he's had. I honestly don't think he would have been half as successful if he had tried to make peace with people. He would have been thwarted repeatedly."

All of this led to that afternoon in the White Horse Pub. As the scotch and ale flowed, the group discussed what Tony Knap called Craig's "big idea," and how to make it happen.

Knap, a respected scientist in the microbiology community, caused some raised eyebrows himself as skeptical traditionalists in Bermuda and elsewhere reacted to his support for Craig's "big idea." Knap, however, was convinced that the *Sorcerer II* project would be a win for his program. "I was director of the biological station, so what I wanted to do was to get as much science done as possible. BIOS had researchers with grants to collect and study microbes from Hydrostation 'S' and to run a microbiological time series, but in those days, they were using more archaic technologies than Craig had at his disposal. So a collaboration seemed a no-brainer."

Craig and Knap decided to begin their collaboration by having Craig dispatch a JCVI scientist to Bermuda early in 2003 to work with BIOS and collect ocean samples at Hydrostation "S" as a prelude to *Sorcerer II's* arrival.

Craig sent Jeff Hoffman—who had just arrived from Louisiana to join the JCVI team—to collect samples on board a BIOS vessel called *Weatherbird II*. This 115-foot, 194-ton diesel vessel lacked not only sails but also any claim of being luxurious. But it was outfitted with advanced labs, instruments, microscopes, glassware, sensors, pumps, filters, and computers—everything oceanographers needed to study the organisms of the sea, and the sea itself.

"We collected two samples on *Weatherbird,*" remembered Hoffman, on two different trips. They would later be added to the first sample taken by *Sorcerer II* to be analyzed and sequenced at JCVI and included in the main paper published about the Sargasso findings in the spring of 2004.

⋎

BACK ON *SORCERER II* ON THAT MUGGY DAY IN MAY 2003, Jeff Hoffman and the team secured the pump at a depth of about ten feet into the Sargasso Sea and then flipped it on. Slowly the seawater was pumped up through a plastic tube and into fifty-liter plastic containers mounted in the stern. "On that first trip, our equipment wasn't all that great," remembered Hoffman. "It worked, but it was a pain." As the containers filled up, the team began running the water through several sizes of fine-mesh filters—with a prescreening that eliminated anything larger than twenty microns across. (Humans can see objects down to about twenty-five to fifty microns; a human hair is about seventy to eighty microns wide). The samples were then run through three filters. The first and largest size captured anything larger than three microns across. The second captured microbes and other material larger than 0.8 microns. The third captured the small bacteria and viruses—everything greater than 0.1 micron. The scientists also deployed gauges and sensors to measure the temperature, pH, salinity, oxygenation, and other characteristics of this tiny patch of sea.

The dock of the Bermuda Biological Station for Research (since renamed the Bermuda Institute of Ocean Sciences), January 2003.

Hoffman, as a desert guy being deployed to work on the sea, later remembered being both excited and anxious. "We were doing something that didn't look like much," he said, "sucking up a bunch of seawater. I mean, who cares? Except that we weren't just scooping up ocean water. We were there to test an idea that maybe there was a lot more life in the ocean than people thought."

Karla Heidelberg recalls that, even before she and Hoffman finished collecting the samples, Craig and Tony Knap headed down to *Sorcerer II's* main cabin to begin celebrating the taking of Sample #1. "This was absolutely a team effort," she said, "but it still makes me laugh that, while we were in the back of boat trying to get the samples—and it was hot, and we were strug-

Jeff Hoffman collecting samples on the
back of *Sorcerer II.*

gling with trying to get everything right, to get stuff over the
side and bring in the water—we heard from down below a big
toast with wine. And they were saying, 'To science! Science is
great, and science is hard!'" She laughed. "They were having a
great time. But hey, we were, too."

As the afternoon wore on, the team finished pumping the
seawater through the filters and pulled in the tubes. This was
the signal Charlie Howard needed to start the process of putting
back up the vessel's sails and turning *Sorcerer II* back toward
St. George's Harbor. That night, Hoffman would bag the filters
containing the microbes and freeze them in *Sorcerer's* minus-80

Celsius freezer, so they could be flown back to JCVI in Rockville for the next stage of the effort—sequencing, identifying, and analyzing everything that had been scooped up from the Sargasso Sea.

Recalling all this years later, Jeff Hoffman smiled. "When we made port and headed back to the bar, we had no idea how many microbes were there, or what they were, or what they did."

But they were about to find out.

2

Planet Microbe

The God Particle.
—LEON M. LEDERMAN

AS CRAIG AND THE TEAM PREPARED TO convey the frozen, microbe-lathered filters from Bermuda back to JCVI in Maryland, what exactly *were* these organisms they were so meticulously collecting and analyzing? And how critical are they to life on Earth, including human life? Mostly, they were bacteria, although the team was collecting other microorganisms, too: viruses, archaea, fungi, algae, and protozoa (microscopic animals). But bacteria—those hidden yet vitally important organisms that connect and sustain all living things on our planet—were the major focus. Together, these tiny life forms represent the final and ultimate character in this book, acting together as a kind of planetary super-organism even as they live their million trillion trillion separate lives as highly diverse, individual cells.

Bacteria and other microbes, invisible to us, exist on our small planet in numbers greater than all the stars in universe. Scientists have found them living everywhere they have looked: in volcanic vents, surviving temperatures that may approach 200 degrees Fahrenheit, and in mines and caverns that penetrate miles into the Earth. They permeate the soil and all bodies of water, from rain puddles to the oceans that cover 70 percent of the Earth's surface. They live in and on every human, hamster, worm, and mayfly, and in and on every plant on Earth.

Some bacteria live in environments so inhospitable—at least to us—that scientists call them "extremophiles." These include the likes of *Pyrococcus furiosus,* thriving in hydrothermal vents deep in the ocean; certain forms of *Synechococcus lividus,* encased in ice as cold as minus-25 Celsius; and the freakishly hardy *Deinococcus radiodurans,* which can survive in so many harsh environments that they're called polyextremophiles.[1] In 2020, Japanese scientists reported that *D. radiodurans* strapped onto the outer hull of the International Space Station survived for three years in the vacuum of space and being exposed to intense ultraviolet radiation from the sun.[2] The *Guinness Book of World Records* calls it "the world's most radiation resistant microbe."[3]

Bacteria called endospores—which consist only of a simple genome, a small quantity of cytoplasm, and thick walls that resist heat, radiation, and other extreme conditions—can lie dormant for millennia and still be revived.[4] In 1995, *Science* reported the revival of bacterial spores embedded in insects preserved in amber for more than twenty-five million years.[5]

Bacteria and other microbes also live in the Earth's atmosphere, which is inhabited by around 10^{22} microbial cells.[6] They get into the air, explained atmospheric chemist Kimberly Prather, "by being propelled from the ground and from the sea by the wind." Much of Prather's work at the Scripps Institution of Oceanography and UC San Diego focuses on how microbes contribute to the ice crystal formation in the upper atmosphere that produces rain. On a sunny afternoon in her office overlooking the ocean in La Jolla, she explained that "at the center of every ice crystal is a particle—an aerosol particle." That is a crucial fact to understand because "if you don't have these particles for the water vapor in the upper atmosphere to form ice on, the water stays vapor, and you have no rain." But these particles "can be dust and other materials—or they can be microbes." On rainy days, in other words, some of the droplets that fall onto faces, umbrellas, flowers, and windows contain microbes.[7] For Prather, "microbes are interesting because they seem to be the best particles to use in seeding clouds to get them to rain." The implication is obvious for cloud-seeding experts like her, trying to figure out how to get more rain to fall on places suffering from droughts. One solution, she said, is to infuse the atmosphere with bacteria.

Most microbes don't stray much, if ever, from the microniche they call home. Others travel thousands of miles, conveyed by winds, waterways, and animals on the land, in the air, and in rivers and oceans. It's not unusual to find that microbes originating in, say, the Sahara Desert have been picked up by trade winds and blown west to settle onto the Caribbean, the southeastern United States, and as far north as Maine.[8] These far-flung movements are assisted by humans—for example, when ships transport ballast water, perhaps taking it in off the coast of Vietnam and later dumping it into San Diego Harbor. Other microbes start in one organism and jump to another, as probably happened in 2019, when a mutated coronavirus, originating in bats in the forests of southern China, infected a person, then proceeded to spread like wildfire from human to human.

As mentioned above, bacteria have even been found in space. Humans have brought them to the International Space Station (ISS) on their own bodies and on equipment and clothing, in water and food, and on mice and other animals and plants flown into orbit for experiments.[9] Former JCVI microbiology fellow Aubrie O'Rourke recently studied bacterial samples from the space station, in part to assess how solar and cosmic radiation in space had affected them. "You have these large particles of radiation just bombarding DNA," said O'Rourke, "and it's happening to the astronauts, and to bacteria on the station as well. So, we're putting up in the Space Station some experiments where we're able to control for radiation and what the effect is." His team is also studying the effects of microgravity, and changes in bacterial species found in astronauts after spending time on the Space Station.[10]

Craig adds that bacteria probably exist beyond Earth—on Mars and possibly on other planets, although we have not yet found evidence of this. "Bacteria are so ancient and efficient," he said, "that it's likely they exist if the same basic ingredients of life exist on other planets."

⋎

BACTERIA COME IN AN astonishing variety of sizes, shapes, and colors, and are usually single-celled organisms, as opposed to being made up, like us, of one hundred trillion cells working together. Still, almost anything goes in the different nooks and crannies of nature, where bacteria sometimes cluster by pairs or form extended chains or mats of cells called biofilms. They range in diameter from around one to two microns and in length from five to ten microns, give or take. (Human cells range from this size up to 150 microns.) While most bacteria are spherical or rod-shaped, some look like commas, spirals, curved rods, and stars.

Individual cells containing DNA that is not housed in a protective nucleus are called prokaryotes. Cells with a nucleus, as found in all mammals, are called eukaryotes. Either way, bacterial DNA is typically arranged in linear or circular-shaped chromosomes that float in the organism's cytoplasm inside the cell wall or membrane.

Different species of bacteria derive energy from various sources. For instance, photosynthetic bacteria use the energy of the sun combined with carbon dioxide and water to produce glucose and oxygen. Other bacteria feed on waste and dead organisms or obtain energy by breaking down chemical compounds in their environment or from organisms like plants or humans that they live on or inside of. Some of the latter live in a mutually beneficial relationship with their hosts, others are pathogenic and harm their hosts.

Bacteria first appeared on Earth billions of years ago. Scientists have discovered fossils of these tiny organisms in Australian Apex Chert rocks that date back 3.5 billion years, and evidence exists that bacteria may have existed even 4.41 billion years ago, not long after the oceans first appeared.[11] No one knows how they came into existence, although we do know that when it began, the Earth was very different than the world we know today.

Visitors from another world arriving three billion years ago would have found a planet devoid of macro life—no fish, reptiles, mammals, birds, or plants. The visitors would have found a thinner atmosphere made up mostly of carbon and nitrogen, and seen continents and oceans that would have seemed empty. But if they had the vision or instruments to detect microbes, they would have found a world teeming with life.

Jumping ahead to about one billion years ago, a visitor would have arrived on a world still inundated with bacteria and devoid of most other life forms. Viruses were probably present by then, although when they first appeared is unknown. One theory has them evolving, when the Earth was young, from a common ancestor that also led to bacteria. But instead of developing more complexity these proto-viruses shed genes and simplified, becoming true viruses around 1.5 billion years ago. This contention has some support in the fossil record. The theory was also tested by applying a sophisticated algorithm built from data on the known evolution of modern proteins critical to viral function.[12]

This second extraterrestrial visitor would also have encountered a hugely changed atmosphere, thanks mostly to the countless photosynthetic bacteria that for over a billion years had been secreting oxygen as a byproduct of their metabolisms.[13]

This shift from a mostly carbon and nitrogen atmosphere to oxygenated air would gradually terraform Earth into a planet that could support the multicellular organisms familiar to us now. They began appearing around six hundred million years ago and eventually included us.[14]

"In deep geological time, microbes were responsible for geo-engineering our planet such that it would eventually be habitable for us, and they continue to do so," said Chris Dupont, who researches microbial and environmental genomics at JCVI. Today, bacteria and other microbes in the oceans produce at least half the oxygen on Earth, and possibly as much as 80 percent.[15] One photosynthetic family of bacteria alone, *Prochlorococcus,* produces upwards of 20 percent of all oxygen in the biosphere of our planet. That's more than all the rainforests combined—plants being the other major source of oxygen in our atmosphere. This process is part of a careful balance that human activity is beginning to alter in ways that are not beneficial to the ecosystem that supports us—a topic to which we'll return in pages to come.

Bacteria reproduce mostly through binary fission: they divide into two "daughter" cells that are identical to the parent cell. Sometimes, however, errors occur in the duplication process. Most of these errors are neutral, but some undermine the survival of a specific bacterium, leading to its death or, if the detrimental mutation is widespread enough in a species of bacteria, to its extinction. Other mutations are advantageous, and organisms live to pass them down to their progeny. Some bacteria pass on genes not through binary fusion, but through close contact with other bacteria. A predatory bacterium may even, by feeding on another bacterium, acquire some of its victim's DNA.

Bacteria reproduce on a radically different timescale than macro organisms like humans. "Most bacteria are dividing multiple times a day," said Chris Dupont. "They're going through tens of thousands of lifetimes themselves in a human lifespan. What we think of as a short time is, for them, probably two hundred generations. So, when you say that they mutate fast, it's partially because of our kind of anthropomorphic viewpoint in how we frame time."

Besides bacteria, as previously noted, two other types of cells are known to exist on Earth: archaea and eukaryotes. The archaea were discovered in the 1970s by microbiologist Carl Woese using the 16S rRNA gene-tagging system to identify different microbial species. Archaea are similar to bacteria, but they have a different evolutionary history and produce energy and perform other functions in ways more like eukaryotes.

We humans are made up of that third basic type of cells, the eukaryotic kind which contain a nucleus and organelles and are enclosed by a plasma membrane. Other organisms made up of eukaryotic cells include all plants, animals, fungi, and protists, and most algae. Although it's difficult to know for sure when eukaryotes first appeared as tiny, microscopic organisms, scientists theorize it was some 1.5 billion years ago.[16]

⋎

THE NET EFFECT of this long sweep of microbial evolution and billions of years of interconnectedness of life has been something that shouldn't have happened in a universe where one of the basic principles of physics is that systems tend toward increasing entropy, declining gradually (or quickly) into disorder. On Earth, against all odds, life has managed to maintain a kind of super-order for billions of years, keeping up a planetary equilibrium that has not only sustained life, but created *more* complexity, not less. "One of the most remarkable properties of life is this ability to create order: to hone a complex and ordered body from the chemical mayhem of our surroundings," wrote Craig in *Life at the Speed of Light.* On its face, this would seem to be "a miracle that defies the gloomy second law of thermodynamics."[17]

How this equilibrium and the maintenance of order works, and has somehow survived for billions of years, is one of the key questions that Craig and his team hoped to make progress toward answering through the Sargasso Sea and global sampling projects. "I have done all this work with microbes to understand how life works," he said, "and to understand the many ways that nature has figured out how to deal with the basics of maintaining life for billions of years. Finding food, and producing and storing energy, and evolving and adapting to new niches—how they do this is the secret of life on Earth, and probably beyond Earth."

"The underlying truth is that this is purely a struggle of life to replicate itself," said Jack Gilbert. "So, if you will, that's evolution." Gilbert thought the novelist Michael Crichton, author of *Jurassic Park,* had expressed it best when he wrote that "life will find a way."[18] "That's the point of equilibrium," said Gilbert. "These weird little pieces of nucleic acid desperately want to survive to be replicated."

Gilbert added that what he'd just said was an anthropomorphic description of this dynamic. "Of course, there is no human drive behind this process," he said. "This is just chemicals striving to be replicated. And the very point of that system is that it either collapses or it persists. And if it persists, it must be continually straining toward an equilibrium that allows that system to persist. Otherwise it collapses and dies."

Craig thinks one of the keys might be that the basics were established billions of years ago, and that all developments since— all the genetics that scientists work to tease out using shotgun sequencing, artificial intelligence, and hopefully one day, quantum computing—have been minor variations. "We have gone on to sequence and analyze hundreds of species, including the human genome," said Craig, "and we can start to understand now the genetic components, the minor changes that have taken place over billions of years, have literally led to us to being able to conduct this interview, to conduct this expedition, to have all the things associated with human life."

ON MAY 13, 2003, as Craig and his team prepared to depart Bermuda after taking those first samples, no one was yet thinking about universal theories or grand hypotheses. Nor did anyone guess that some twenty years later the science of microbiology, which in 2003 was practiced mostly through small-scale experiments and with limited understanding of the diversity of microbes in the sea, would be seriously pondering universal theories and trying to grasp the world of microbes on a planetary scale. "The idea among scientists of thinking big about microbiology started in large part with Craig and the expedition to the Sargasso Sea," said Jack Gilbert. "His idea of going global, and bringing in the technology to make this possible, changed things."

But this would come later. Back in May 2003, the goal was for Craig and Jeff Hoffman to convey the frozen, filtered microbes from Bermuda back to Rockville, Maryland.

They arrived at Bermuda's international airport a few days after *Sorcerer II* returned from the Sargasso Sea, with Hoffman toting a cardboard box containing an insulating core filled with dry ice and three 142-millimeter filters folded into thirds. Hoffman was flying in a private jet for the first time and was being initiated into a newbie ritual that placed him in a side seat in the plane by the door, right next to the bar. "The new guy had to bartend," he said.

Craig was there, he remembered, and a couple of others, drinking beers as the plane taxied and they took off, carrying their treasure trove from the Sargasso Sea. They were headed toward Maryland, where these microbes in their billions would be sequenced and analyzed—coaxed to give up some of the secrets that lay hidden in their DNA.

3

The Ocean's Genome Goes Meta

You can overinterpret DNA sequence data, but if you're careful, you can use them as a clue.

—CRAIG VENTER

ON A CLOUDY MAY AFTERNOON IN 2003, with thunderstorms threatening, the private jet carrying Craig Venter and Jeff Hoffman from Bermuda landed at the Montgomery County Airpark in Gaithersburg, MD. From there, Hoffman conveyed the frozen filters to a nondescript lab building in nearby Rockville, a temporary facility where scientists would begin to prepare the DNA from the Sargasso Sea microbes for sequencing and identification. At the time, Craig's various institutes were housed in separate locations but a new campus to bring all their operations under one umbrella was being completed in Rockville. The Institute for Genomic Research (TIGR), his flagship research organization, had rented this lab as temporary space in the meantime.

The core of Craig's operations in Rockville centered on his institute's "next gen" sequencers. Mostly, these were ABI 3730xl DNA Analyzers from Applied Biosystems, each machine roughly the size of a mini fridge and costing around $300,000. Dozens of them were lined up in rows on tabletops in several large rooms in Rockville. "They could do ninety-six sequences at the same time instead of just one," said Hoffman, which was speedy at the time. The sequencers worked by tagging each genetic letter—every A, G, C, and T—with a specific fluorescent dye that could be read by a laser and recorded as code by a computer.

All this sequencing heft was crucial to what Craig was setting out to do with the Sargasso Sea microbes—something that had never been tried at this scale. This was to use shotgun sequencing to go beyond sequencing a single species or organism's DNA, which had been the norm, including for the Human Genome Project.

Instead, Craig was proposing to sequence and reassemble the thousands of individual organisms and species contained in the heterogenous samples from the Sargasso Sea—an idea that many scientists thought would fail. This meant reinventing and revising every step of the sequencing process, starting with the way lab techs cracked open and isolated DNA from so many individual organisms all at once. A challenge was how to make sequencing libraries that could accommodate a large number of different microbes at the same time while still keeping the randomness of the sequences. "At the time," Craig said, in an understatement, "this was not an easy task."

To prepare the raw samples for sequencing, Hoffman turned to Cyndi Pfannkoch, a lab tech at JCVI who had worked previously on making the DNA ready for sequencing on the human genome project and was now assigned to the temporary lab in Rockville. "That's where we started working on the Sargasso Sea," remembered Pfannkoch, "and where we developed how we were going to filter things. We had two or three labs, I think, that were kind of like the NIH in the fifties. That sort of decor. The black Formica benches with the drab green walls, with a couple little benches like tables on the side."

Trained as a soil microbiologist, Hoffman adapted techniques for extracting DNA that he had learned in the deserts of Arizona while pursuing his PhD. "This was pretty standard," he said, "although I had never used them on microbes from the ocean." Hoffman, Pfannkoch, and a small team at the institute began by taking the frozen, microbe-containing filters brought back from Bermuda and cutting them up into tiny pieces. They then used special enzymes designed to break open the cells and extract the DNA. "Turns out that ocean microbes break open pretty easily," said Jeff, "compared to desert soil microbes that have thicker membranes." The full extraction process took about three days.

"In part, we relied on Jeff's prior background," recalled Craig. "You had to be able to break open all the cells. So they tried some different experimental protocols to get out all the DNA.

The microbiologists on the team had experience breaking open cells and Jeff had experience with desert soil microbes. Some cells from soil are extremely hard to break open because they form these little microcapsules. Each member of the team brought forward various protocols that would break open all the cells and give us the highest DNA yield."

The next steps were to create multiple copies of the DNA in the samples and then to literally break up the DNA into fragments of anywhere from five hundred to two thousand base pairs, using a machine called a *nebulizer*. The fragments were then run through one of those Applied Biosystems sequencers, which tagged and identified each genetic letter.

Once the sequencing was complete and the computer processing of files was done, a team of computational biologists at the institute took the digital sequences and went to work trying to reassemble these fragments of code into the organism's naturally occurring chromosomes. The team did this by looking for overlaps in the code of different fragments. A detailed description of how this is done is provided by the National Center for Biotechnology Information in its *NCBI Handbook*.[1] But for simplicity let's say that, after the nebulizer had done its blasting, the fragments left by it included the following three:

Fragment 1: —————TCATGCTTGAC—————TACAGC

Fragment 2: TGCATCATGC—————GCTATACAGC

Fragment 3: —————TTGACGCGGCTATAC———

A computer quickly identifies the overlapping parts of these fragments and is able to reassemble the full sequence they cover:

TGCATCATGCTTGACGCGGCTATACAGC

This is a process that can work when multiple copies of a genome of a given organism have been gathered and shattered to yield their different fragments. The more copies, the greater the odds that stretches of identical sequences of DNA will be found and that random fragments can be compiled by the computers into contigs.

An even simpler way to think of this is to imagine that someone has printed several copies of, say, a *New York Times* article and sliced them up in different ways, yielding many isolated strings of characters. The fragments make no sense until you start detecting sequences in different clippings that match up exactly. Let's take, for example, the first sentence of an article written in 2001 about the human genome project by Nicholas Wade.[2] Here are some possible fragments of it:

/ e long-held beliefs about human biology. /
/ The publication of the first interpretat /
/ quence this week overturns some long-held beliefs abou /
/ terpretations of the human genome sequence th /

And here are those fragments stitched together to reveal the original sentence:

The publication of the first interpretation of the human genome sequence this week overturns some long-held beliefs about human biology.

⋎

WHOLE GENOME shotgun sequencing was first developed in the mid-1990s when Craig and Hamilton "Ham" Smith—a Nobel laureate and Craig's close friend and collaborator at TIGR, Celera, and JCVI—invented the process for sequencing *Haemophilus influenzae.*

"Others had used the term 'shotgun sequencing' to describe what they were doing, which is confusing," said Craig. "For example, Fred Sanger in 1977 used this term for the methods he used to sequence a virus for the first time, the Phi-X174."[3] So did a team at the University of California at Davis in 1981, who used the same approach as Sanger when they sequenced the Cauliflower Mosaic virus.[4] But both viruses were prepared for sequencing not with nebulizers shattering DNA into small, random fragments, but by using a more traditional method that deployed restriction enzymes. These work like chemical scissors to cut up DNA into pieces at precise locations in a genetic sequence. The pieces are then sequenced one by one and manually reconnected in a computer. In fact, it was Ham Smith along with two others who pioneered the use of restriction enzymes—a discovery that earned them the 1978 Nobel Prize in Medicine.[5]

"Instead of restriction enzymes," said Craig, "we randomly sheared the DNA into small fragments, twenty-five thousand of them in a single tube for the first genome. Then all these fragments were sequenced, and the twenty-five thousand fragments accurately reassembled." Craig wrote about this process for sequencing *Haemophilus* in *Life at the Speed of Light:* "The result was that the 1.8 million base pairs of the [*Haemophilus*] genome were re-created in the computer in the correct order. The next step was to interpret the genome and to identify all its component genes."[6]

In 1995, the team wrote up the results of the *Haemophilus* sequencing and its interpretation in a *Science* article titled "Whole-Genome Random Sequencing and Assembly of *Haemophilus influenzae Rd.*"[7] "The fact that we could assemble *Haemophilus* with an algorithm so fast and so accurately defied everybody's expectations," remembered Craig. "They had used the same argument they were using at the time for sequencing the human genome, that it would take decades to sequence that much DNA using the old methods. But then we succeeded with *Haemophilus*, proving that it was mathematically possible to do it much faster. It also proved you could use this method to sequence the human genome. This wouldn't have been possible without doing *Haemophilus* first."

When the *Haemophilus* genome was published, he added, "Fred Sanger even sent me a nice handwritten note ... saying he always believed my approach would work, but he never got the chance to test it because his colleagues each wanted their own piece of DNA."[8]

After sequencing *Haemophilus influenzae* and *Mycoplasma genitalium*, Craig received a great deal of attention in the media and in scientific circles. Ham and Craig were invited to give the president's lecture at the American Society of Microbiology at its annual meeting in Washington, DC. "Ham introduced me, and I gave the lecture," said Craig. "At the end, a very rare event in science happened: twenty-thousand-odd scientists rose to their feet and gave us a standing ovation for sequencing the first organism in history."

"This success allowed us to get major funding and to keep going with these early experiments and validations of shotgun sequencing using bacteria," he said. One of the primary funders, supporters, and cheerleaders of almost all of Craig's projects after *Haemophilus* was Ari Patrinos, who in 1995 was put in charge of biological and environmental research at the Office of Science in the US Department of Energy (DOE). Patrinos later recalled how he missed the chance to fund *Haemophilus* because of all the flak he got from reviewers at the DOE about the proposed project. "I had wanted to fund the sequencing of the first microbial genome," he said, "but the reviews of the project all came back negative. All these experts didn't think Craig could do it. I eventually overruled them, which I was allowed to do, but I had to go through the paperwork to get it overruled—and meanwhile Craig got some private money and managed to fund that first one without DOE, because it took too long for me to get the permission." His office "funded pretty much everything after that," however. "And it was one of the best things I did," Patrinos said, because Craig's work "was groundbreaking, and very much changed the mindset of the scientific community in that field."

Craig remembered a revelatory finding from the early days of the DOE's support. "After *Haemophilus* and *M. genitalium,* the DOE came in and wanted to fund us to do twenty or thirty genomes," said Craig, "and they set up an advisory committee to help us choose: What are the most important microbial species on the planet that we should do first?" The committee included famed microbiologist Rita Colwell, who would go on to serve as director of the National Science Foundation, and Carl Woese, the microbiologist who developed the 16S rRNA process for barcoding species and discovered the existence of the archaea cell type.

Having Colwell on the committee suggested an obvious candidate, because "the species of her life's work was cholera." As Craig recalled, "I thought this would be interesting to sequence next because there was this big debate in the cholera community whether cholera had one or two chromosomes." Carl Woese, how-

Ari Patrinos in Greece.

ever, was sure that there was no need to use the shotgun method on cholera because its 16S rRNA code was a close match with another bacterium, *E. coli*, which has only one chromosome. But "one of the things I like best about shotgun sequencing is that it's hypothesis-free," Craig continued. "It doesn't matter what your belief was before—whether it was one chromosome or two chromosomes, or that the 16S gene is similar to *E. coli*. It's like a truth machine, because if there's independent elements in there, it assembles them independently, as they really are, instead of what people think they might be."

Sure enough, the truth came that cholera has two chromosomes.[9] "One of the chromosomes had a 16S rRNA tag in it, and that chromosome looked a lot like *E. coli*," Craig said, which meant that, at best, "Carl Woese was half-right—because the second chromosome didn't resemble *E. coli* at all, and probably came from a fusion of two bacteria at some ancient time."

Considering what organism to sequence next, Craig wanted to focus on one that exists in extreme conditions, hoping for insights into how it is able to survive. As he later wrote: "In 1996 we purposely chose an unusual species for our third genome effort: *Methanococcus jannaschii*. This single-cell organism lives in an extraordinary environment, a hydrothermal vent where hot, mineral-rich liquid billows out of the deep seabed. In these hellish conditions the cells withstand over 245 atmospheres— equivalent to the crushing pressure of 3,700 pounds per square inch—and temperatures of around eighty-five degrees centigrade (185 degrees Fahrenheit). That in itself is remarkable, as most proteins denature at around fifty to sixty degrees centigrade, which is why egg white becomes opaque when cooked."[10]

Craig collaborated with Carl Woese on *Methanococcus,* the first archaea cell type ever to be sequenced.[11] "The sequence did not disappoint," he wrote. "The *Methanococcus* genome broadened our view of biology and the gene pool of our planet. Almost 60 percent of the *Methanococcus* genes were new to science and of unknown function; only 44 percent of the genes resembled anything that had been previously characterized. Some of *Methanococcus's* genes, including those associated with basic energy metabolism, did resemble those from the bacterial branch of life. However, in stark contrast, many of its genes, including those associated with information processing, and with gene and chromosome replication, had their best matches with eukaryote genes, including some from humans and yeast." The *Methanococcus* genome study appeared on the front pages of major newspapers "and generated some interesting headlines," noted Craig.[12]

The process also inspired people to imagine how life might have begun on the early, volcanic Earth, and how life might exist on other worlds where temperatures and the presence of caustic chemicals had previously seemed to preclude it.[13]

The next genome Craig and his team sequenced was another extremophile, called *Archaeoglobus,* which lives in oil deposits and hot springs. "The organism uses sulfate as its energy source but can eat almost anything," wrote Craig. "Our first analysis of more than two million letters of its genome revealed that one quarter of its genes were of unknown function . . . and another quarter encoded new proteins." By revealing novel biological mechanisms to produce energy, *Archaeoglobus* also had the potential to suggest alternatives to human uses of fossil fuels.

The surprising result of another experiment during this time proved that shotgun sequencing could be used to separate out and identify more than one species of bacteria at the same time from the same sample. This revelation came when Craig's team was asked to identify a strep bacterium that was causing severe pneumonia in a patient in Norway. The shotgun process revealed that the culprit was actually two different bacteria. "That's probably why it was such a bad infection," said Craig. The puzzle was solved when "the assembly spit out two independent, closely related genomes, where people thought there was one. This is what convinced me from that point on that each species's genome on this planet has a unique mathematical solution." Along with this conviction came another one. "It also made me certain that we could sequence mixed populations of bacteria"—even as large a population as might be present in, say, two hundred liters of seawater from the Sargasso Sea.

This, of course, was a prelude to Craig's outrageous idea in 2003 to use shotgun sequencing to make sense of samples containing multiple species of microbes. And as revolutionary as that work was, it was also not lost on Craig and others, even in the late 1990s, that shotgun sequencing might also allow scientists to sequence in a single sample *all* the bacteria in that Sargasso Sea sample—or inside a deep well, volcanic vent, or human gut.

The idea of sequencing all of the DNA in all of the microbes in a specific ecosystem or location and studying the results would launch a new field called *metagenomics.* This term has been used to describe studies of everything from specific patches of the ocean to the ultimate metagenomic aspiration—to sequence all the microbial DNA on our planet, every last A, G, C, and T in every microbe. This would reveal the base code of the vast network of microbes that undergird life on Earth. Craig once told a reporter that he wanted to sequence all the microbes on Earth. He was mostly kidding, but not entirely.

ON MARCH 4, 2004, Craig announced the key results reported in the Sargasso Sea paper at a press conference in Washington, DC, a month before the study was published in *Science* on April 2, with Karin Remington as the lead bioinformatician.[14]

By then, Craig's around-the-world expedition was well underway, and *Sorcerer II* had reached the Galapagos Islands. Craig, tan and fit, had flown back to Washington, DC, to tell a

Karin A. Remington on *Sorcerer II.* She was the lead bioinformatics scientist on the first ocean samples.

crowd of reporters assembled at the National Press Club that his team had discovered nearly two thousand different bacterial species, including 148 that had never before been seen, and some 1.2 million new genes.[15] Conservatively, that doubled the number of genes previously known from all species, micro and macro, in the entire world.[16] It was a DNA haul showing the microbial diversity and sheer numbers of bacteria in the Sargasso Sea to be far, far greater than anyone had imagined. "Now, with these new tools," he announced, "we can see what everyone has missed to date, which is the vast majority of life."[17]

Tony Knap later recalled the general response to the news: "I would have to say there was a little bit of jealousy, because everyone wanted to be the first to do this and then Craig did it. People had been working on similar projects, but no one had published anything. And then this paper sort of blew people out of the water. There were a lot of rumors that there couldn't be that much diversity in the ocean, that we must have been sampling the sewage."

"This paper turned out to be *the* paper for the ocean microbiome," Knap continued. "People now acknowledge it as a pioneering effort in oceanography. Of course, you can do these many times over today with newer equipment. But back then, this was incredible, and very expensive. We were able to do it because I had access to the marine samples, and he had access to the sequencers and the science. When I talk to people about that paper, no matter what field in oceanography, everybody knows that paper and the impact it had on the field."

"The total number of genes they found is mind-boggling," Paul Falkowski, an oceanographer at Rutgers University in New Brunswick, New Jersey, told a *Genome News Network* reporter.[18] When Andrew Pollack of the *New York Times* reported on the study, he reached out to Stephen Giovannoni, a professor at Oregon State University who studied microbes in the Sargasso Sea. Giovannoni said, "He's allowed us for the first time to see that diversity." The same article quotes David M. Karl, a professor of oceanography at the University of Hawaii, saying "It's almost hyperbole. How can you imagine one sample having a million new genes and proteins that we don't know anything about?"[19] Pollack talked to several oceanographers who said they didn't learn much that was new, although the study's findings "confirmed the variety of ocean life and provided a giant parts list that would be studied for years to come." Others complained that Craig was not collecting enough ancillary data such as water temperature and salinity. They also said that some of the bacteria he found might have come from contamination of the sample, though the *Science* article said the researchers guarded against contamination."[20]

"We demonstrate here," concluded the study itself, written in the usual careful, understated style of a scientific journal article, "that shotgun sequencing provides a wealth of phylogenetic [evolutionarily related] markers that can be used to assess the phylogenetic diversity of a sample with more power than conventional PCR-based rRNA studies allow. We find that although the qualitative picture that emerges is similar to that based on analysis of rRNA genes alone, the quantitative picture is significantly different for certain taxonomic groups. Further, just as shotgun sequencing provides a relatively unbiased way to examine species diversity, it can also allow a relatively unbiased identification of the diversity of genes in particular gene families."[21]

The study described the discovery of a large number of certain species of bacteria—including *Burkholderia* and *Shewanella oneidensis*—that are normally found on land or in nutrient-rich bodies of water. Also found were bacteria known to be common in the ocean, including *Prochlorococcus* and SAR86. "The team found nearly 800 new genes for proteins sensitive to light, suggesting that more bacteria than were previously thought might be converting light into energy for the organism other types of energy," wrote Pollack.[22] "This is hundreds more than have been found before, and IBEA researchers want to study these genes further to explore the possibility of producing hydrogen as a fuel source.[23]

"The Sargasso project was one of the key drivers of what became modern metagenomics," said Chris Dupont, speaking in his office at JCVI in 2019. "It was huge, the discovery of so many new taxonomic lineages that we simply did not know existed, period. In some cases, it's because the universal marker 16S rRNA doesn't account for the differences in some of these lineages, because you get species with the same 16S rRNA marker that behave fundamentally differently. With shotgun sequencing you got a much more complete picture."

"Craig was the right person to do this," said Juan Enriquez. "You need a scientist who understands outliers and serendipity. All the best scientists do. They can say: 'holy shit, this is surprising. Let's pull on this very hard.' He's never tentative when he finds something that is unusual to pull on."

$\curlyvee$

ONE UNEXPECTED REACTION to the Sargasso project and paper came from a smattering of environmentalists and activists who declared that the Sargasso microbes collected on *Sorcerer II* were stolen from Bermuda. There are international treaties governing how foreigners can extract and use valuable natural resources from precious stones and metals to fossil fuels to plants and animals—and microbes—and they suspected Craig's team of having violated these.

At the time, these treaties were relatively new; the most important of them went into effect in late 1993, having been introduced during the United Nations' 1992 Conference on Environment and Development in Rio de Janeiro, and agreed to by 168 nations. Known as the UN Convention on Biological Diversity, this pact was the product of decades of debates and negotiation and included provisions governing who owned the rights to each country's "genetic resources"—any DNA of plants, animals, algae, fungi, and bacteria that might one day lead to a blockbuster drug or a miraculous new chemical or source of energy.[24]

In part, these treaties were a reaction to centuries of resource exploitation by Europeans in their former colonies—an extraction of natural wealth that often began when geologists, botanists, and other scientists accompanied or followed in the wake of explorers and conquerors who laid claim to a territory. This legacy added to a wariness in some countries about what Craig Venter—a famous white scientist who had recently run a multibillion-dollar genomics company—was really doing in places like Bermuda. Was he truly looking for microbes purely for research, and did he really plan to deposit any genes found in public databases, as he claimed? Or was he secretly looking for organisms to profit from?

"Venter's microbe-hunting expedition raises serious unanswered questions about sovereignty over genetic resources and resource privatization through patenting," said a regional director of the ETC Group, a Canadian-based environmental activist organization that issued a public criticism of the *Sorcerer II* project during the summer of 2004. ETC also dubbed Craig the "Biopirate of the Year."[25] If that award entitled him to some kind of plaque, Craig says he never got it. "If I had," he said, "it would be hanging on my wall in my office."

More serious was a *Nature* piece titled "Bermuda Gets Tough Over Resource Collecting."[26] In it, science journalist Rex Dalton reported:

> When Craig Venter steered the *Sorcerer II* into Bermudian waters early last year, he was searching for ocean microbes. And he found plenty...

But his voyage into the Sargasso Sea also took the genomics pioneer into uncharted waters. The rules on bioprospecting in this small British protectorate are still a work in progress. And experience with expeditions such as Venter's has prompted Bermuda to temporarily shut down some research projects until it strengthens its regulations.

"Here we discovered over a million genes in Bermuda's water," remembered Craig, "and the implication in Bermuda was that we *patented* a million genes. But the team did not patent anything. Everything went into the public domain."

For some naysayers, sharing the data publicly wasn't acceptable either, recalled Tony Knap, director of the Bermuda Biological Station for Research. "These people said that we had cheated Bermuda out of opportunities to make revenues, although you can imagine that, if we would have kept it private, we would have been hammered, too." Knap, who also ran afoul of the critics for his collaboration with JCVI, countered Dalton's article with a letter back to the editors of *Nature:*

Bermuda welcomes careful prospectors

Sir:

Despite the claim in your News story "Bermuda gets tough over resource collecting" (Nature 429, 600; 2004), Bermuda's Ministry of the Environment has not shut down any research projects relating to biodiversity access, even on a temporary basis. New laws and regulations are under development to enhance bioprospecting, not to prevent or hinder such research activities.

   Contrary to your News story, the Ministry of the Environment . . . appreciates the station's responsibility in ensuring a proactive and consultative approach to issues of environmental access.[27]

The Bermuda kerfuffle was not the last time the *Sorcerer II* project would run up against sensitivities and challenges on issues of biodiversity and Craig's true motives. During the world trip, securing rights to take samples in the territorial waters of nations along *Sorcerer II's* route became a full-time operation for a small team at JCVI, led by Robert Friedman, a PhD in ecological systems analysis turned bio-diplomat. This usually went smoothly, although there were moments when the expedition faced dicey situations with uncooperative governments and officials. A few times, moves by officials to delay or rescind permissions even led to boardings by armed police. *Sorcerer II* was ordered not to leave port while authorities sorted out the politics of whether Craig's team would be allowed to collect samples and send them back to JCVI in Maryland. "Usually, everything turned out okay," said Craig, "but we had some close calls."

CRAIG WRAPPED UP the press conference in Washington, DC, by announcing that he and his team were already deep into a global voyage, having launched from Halifax in Canada some six months earlier, and had a crew at that very moment collecting samples in the Pacific Ocean. Grinning, he told the journalists he had just flown in from the Galapagos, where the ship's crew had taken some of the most fascinating samples yet—drawing microbes from mangrove swamps, freshwater ponds, sulfuric vents, and more.

Mentioning the Galapagos also gave Craig the opportunity to talk about one of his great inspirations for launching a globe-circling expedition: Charles Darwin's voyage on HMS *Beagle*, which had stopped in the Galapagos 168 years before *Sorcerer II* arrived. Craig often talked about how he and his team, like

Darwin, had embarked on an epic search for new lifeforms—but micro, not macro.

This sort of talk was already leading to some eye rolling as Craig seemed to be comparing himself to the greatest biologist in history. It would lead to headlines like the 2004 *Wired* cover that read: "He cracked the genome. Now he's chasing Darwin."[28] Yet even in the summer of 2004 it was becoming clear that Craig would beat Darwin in at least one respect: the sheer quantity of new life his team was pulling out of the sea.

"I will be joining the vessel very soon to head to French Polynesia," Craig told the reporters as he ended the press conference. "It's tough duty."[29]

PART II

THE VOYAGES

4

Halifax to the Galapagos

As a kicker, Venter also revealed that the Sargasso trip was
only a pilot project for a vastly more ambitious undertaking:
His yacht the Sorcerer II was at that moment in the Galapagos
Islands, embarked on a two-year, round-the-world expedition
that promised to overwhelm the huge amount of data
from the Sargasso Sea.

—JAMES SHREEVE, *WIRED*

SORCERER II'S CIRCUMNAVIGATION OF THE PLANET began on an overcast Sunday in August 2003, when the ship sailed under gray swirling clouds into Canada's Halifax Harbor. The crew had navigated there from Nantucket, where the ship had been anchored since arriving from Bermuda and the Sargasso Sea three months earlier. Craig chose Halifax to start the expedition in part because its far-north waters would undoubtedly contain different microbial communities than the warmer seas they would sample for most of the global trip. But there was another reason that was less about science than about history—and even had a touch of romance. Craig was channeling an earlier voyage of scientific discovery that had inspired him—the circumnavigation of the Earth in the 1870s by a two-hundred-foot, three-masted ship called HMS *Challenger,* which visited Halifax in 1873.

Like Darwin and the *Beagle* earlier in the nineteenth century, *Challenger* had a mission to seek out and take samples of new life, but in its case, not from land. The scientists on board this warship-turned-research-vessel were, for the first time in history, searching for life deep below the surface of the ocean. In part, this was to test the then-prevalent theory that life could not exist deeper than three hundred fathoms (eighteen hundred feet)—in other words, that all water and seabed beyond that depth constituted an "azoic zone."

To Craig, this theory, first propounded in 1844 and accepted for decades after, sounded a lot like the scientists in the early 2000s who presumed there was less micro life and diversity in oceans than in terrestrial soil and animals. Like the *Challenger's* lead scientist Charles Wyville Thomson, a natural historian from the University of Edinburgh, Craig was setting out to test and, he suspected, to disprove what he considered a false conjecture. He also was taken with how Thomson had deployed the latest in high-tech inventions of his day to delve into a world previously hidden and inaccessible—such as a steam-powered dredging system, with hemp ropes 181 miles long, that could scrape up matter from the ocean floor, even several miles down. Other equipment, like thermometers, barometers, and sounding leads to determine depth, also represented the state of the art, just as *Sorcerer II's* instruments did 130 years later. *Challenger* had had the latest microscopes, which its scientists used to take the first systematic peek into microbes in the deep ocean, although instruments in the 1870s were able to reveal only the largest microorganisms.

It hadn't taken long for *Challenger's* scientists to debunk the theory of the azoic zone, as the ship's great dredge hauled up an astonishing variety of plants, animals, and microbes.[1] Over the course of a four-year expedition, Thomson and his crew conducted 492 deep sea soundings, 133 bottom dredges, and 151 open water trawls, plus 263 serial water temperature observations. They discovered and took home some forty-seven hundred new species.[2]

Images of these samples and the data collected were eventually issued in a fifty-volume report. John Murray, one of the first ever true oceanographers, supervised its publication and wrote most of its text. In its final, summary volume, he described the work of the Challenger scientists as "researches and explorations which mark the greatest advance in the knowledge of our planet since the celebrated geographical discoveries of the fifteenth and sixteenth centuries."[3]

By Halifax, Craig's team already had at least a partial answer, from the Sargasso Sea findings, to the question of how much micro life there might be in the sea. But they needed much more data from sites around the world to prove that prevailing theories in 2003 were wrong—evidence that the scientists would gather using cutting-edge technologies that also measured salinity, oxygen in the water, depth, and temperature, and that would then be analyzed back in Rockville using the latest DNA sequencers, nebulizers, restriction enzymes, advanced computers and algorithms, and much more. The team's nimble, high-tech ship also allowed it to cover great distances with a speed and ease that Thomson and *Challenger's* captain, George Nares, could only have dreamed about.

SOON AFTER ARRIVING IN HALIFAX, Jeff Hoffman and his team collected a sample in the swirling waters of the city's crowded harbor. With the sun setting behind a cluster of modern, high-rise buildings, they prepped the pump system and set up the filters and instruments the scientists would drop overboard to make temperature, salinity, and other measurements key to understanding the environment supporting whatever microbes they found. This was the fourth sampling location, the first having been in the Sargasso Sea and the second and third in the Gulf of Maine on the way to Canada.

Hoffman and the team had made a few changes in the sampling process since Sargasso. "Halifax was the first time we had small filters," designed to catch viruses and other very small microbes, Hoffman explained. "This ended up being a hassle, because the small size made for slow flow rates. The water was also choppy, which meant the equipment kept bouncing around, something we later fixed." The upshot was that the team didn't finish taking that sample until 4:30 AM—"way past my bedtime."

Over the next two days, while Hoffman took more samples, Craig met with local scientists and government officials, which became the pattern every time *Sorcerer II* stopped in a major port. The team also explored this city of about four hundred thousand, where steel and glass high-rises overlay a British colonial town founded in 1749.

Sorcerer II spent a third day in Halifax, although the team did not. Early in the morning of August 23, Craig, Jeff Hoffman, and several others left the ship and drove to the Bay of Fundy to take a sample on the opposite side of Nova Scotia. Collecting water from the bay wasn't easy. Tides there rise and fall up to fifty feet a day, the largest gradient in the world. "To take a sample in such a very rapidly moving tide was like sampling in a fast-moving river," said Craig. "While we were turned down by most, we found a willing, brave fisherman with a small boat who had a large engine that could outrun the tidal currents."

During the Bay of Fundy day trip, Craig visited the famed medical geneticist Victor McKusick, who had a summer house there. Sometimes referred to as the father of medical genetics, McKusick was then eighty-one years old and still a professor of medicine and medical geneticist at Johns Hopkins. He also continued to oversee the Online Mendelian Inheritance in Man, a comprehensive, free database of human genes and phenotypes that he had first begun assembling in the 1960s. The two men had been acquainted since the early 1990s, when Craig was focused on expressed sequence tag (EST) analysis at the NIH. As well as supporting that work, McKusick championed the human genome project at Celera, in both cases "because he believed that the genome could thus be completed sooner," according to his 2008 obituary in *Nature*.[4]

Juan Enriquez, who was with Craig when he visited McKusick, knew the meeting meant a lot to Craig. "He was coming off the intensity of the human genome project," said Enriquez, and "everyone was criticizing him. People were thinking he was crazy to be out there on his sailboat talking about microbes in the sea. Now, here was the father of human genetics giving him his blessing."

McKusick joined Craig and his team to help take a sixth sample from the Bay of Fundy, and then invited the group to drop by his cottage. "There we had a most pleasant surprise," recalled Craig. "Victor's wife, Anne, had baked a fresh blueberry pie from berries she had picked that afternoon. It was most delicious and added a very touching human element to start our expedition. We drove very satisfied back to the boat in Halifax."

⌄

ON AUGUST 24, the ship set sail from Halifax on a foggy morning and headed south, returning to Hyannis, Massachusetts, some four hundred miles to the south. Arriving in the harbor, *Sorcerer II* looked majestic amid the sailboats and power boats in the mooring field, most of them smaller (with the exception of a handful of superyachts). Jeff Hoffman and his team took a seventh sample in Hyannis Harbor. Hoffman also reported a dramatic scene in the water nearby: "I saw a whale getting eaten by a shark."

"No one believed me," he said, "but it happened."

"And we still don't believe you," laughed Heather Kowalski when Hoffman recounted the story years later. Kowalski was head of communications for Craig's institute at the time; they would marry in 2008. She dipped in and out of the global trip as she planned and coordinated Craig's schedule of talks and meetings with local scientists and media.

From Hyannis, *Sorcerer II* made a short sail to Newport, Rhode Island, where Charlie Howard got serious about outfitting the vessel for a journey around the world. "The challenge was to equip *Sorcerer* to be independent of shore support for months at a time across the Pacific," Howard recalled. "To have good navigation gear and communications for anywhere around the world, and to have enough redundancy and spares to be able to keep her going with any breakdowns that might occur. We added satellite, voice, and internet communications with a new stabilized dish antenna."

"We also added all kinds of spare tools, including a sewing machine and a MIG welder, just in case," he added. Other goods included extra line and sails and deep stores of food. At the same time, Howard had security cameras and recorders installed and, adding to the scuba compressor already on board, secured enough diving gear for six people to be able to dive at once.

Then there was the challenge of finding a crew with the right skills and commitment to stick with a long voyage where the going could get tough. After reaching out to friends and associates, Howard found two key crew members almost by accident in Newport: Cyrus Foote, who joined as a deckhand but would later become first mate; and his girlfriend, Brooke Dill, who became a crew member and dive master. Dill also had a keen interest and some training in marine biology. "One day we heard about this scientist looking for crew members," she remembered. "I was excited because I wanted to go blue-water distance, but I wanted it with a purpose. I didn't just want to work on a luxury yacht. My background is in zoology, and I was doing research on humpback whales in Hawaii prior to Newport."

Dill's other job on *Sorcerer II* was to help Hoffman and other onboard scientists collect samples. "We called ourselves Jeff's entourage," she said. Dill also ended up writing some of the log entries during the expedition and helped with planning and keeping track of the route.

"The boat was in Newport for three months," said Hoffman. "I went back to Rockville and loaded up a bunch of gear, like the new filter rigs, and drove them to Newport, and that's where we put the new filter system together. We made an 'arm' to go in the water with the pump to get the sample to the filter from the surface and made some other modifications."

Sailing out of Newport on November 17, *Sorcerer II* headed for Annapolis, Maryland, its final step for making preparations and modifications. It arrived the next day in a dense fog, surrounded by what the ship's log describes as the "heavy traffic" of other vessels. Located about forty-five miles from the JCVI institute in Rockville, the stopover gave easy access to a stream of scientists and technology support people, who came and went as they installed more new equipment, including computers, sonars, instruments, communications gear, and microscopes.

In Annapolis, Hoffman worked with Cyndi Pfannkoch and other institute scientists to further refine the sampling protocol. This included bolting down the previously tipsy filter holders and figuring out how much water they had to pump through the filters. They also honed the process of cracking open, nebulizing, and prepping microbial samples for sequencing after they were flown back to Rockville, with Pfannkoch again heading up the *Sorcerer II* sequencing operation at JCVI.

In Annapolis, Craig, Charlie Howard, and other key orga-
nizers also finalized their route. "Craig decided where we were
going in general terms," Charlie Howard recalled, "say, to Cocos
Island and to the Galapagos, which was a must. French Polynesia was also a must—and the Cook Islands and Fiji. There's a
normal sailing route that people take when they head west around
the world, where the winds and currents are favorable—a general migration route for sailboats and earlier sailing ships that
we generally followed."

Originally, Craig had wanted to do a loop around South America, sailing around Cape Horn and into the Pacific. "But the sea-
sons and currents were not ideal," said Howard. Plan B—the route
they ended up taking—was to transit the Panama Canal and then
continue west into the Pacific.

The other problem with the South American plan was the
near certainty that Brazil's government would deny any request
to take samples in its territorial waters. Suspicions about the
expedition's motives, and concerns over who might profit from
any valuable microbes discovered, would continue to crop up
during the voyage, as Bob Friedman's team at JCVI negotiated
permits to take samples in territorial waters of countries along
the planned route.

One of the challenges in these talks was interpreting the
terms of biodiversity treaties passed in the 1990s and early 2000s
that were a bit fuzzy in places. Specifically for any party wanting
to take a country's natural resources to conduct research, the
rules for gaining permission were not clear.[5] This led to discussions about how to determine the potential worth of invisible
organisms that had in most cases never been characterized and

might or might not have potential commercial applications or value. It was also difficult to say if many of the more common microbes seen throughout the oceans really belonged to a single country.

"One of the ironies of this whole situation," said Robert Friedman, "was that the microbes we're talking about were constantly moving around in the ocean." A bucket of microbes in one nation's territorial water on a Tuesday could by Wednesday "belong" to another nation—or to no nation at all, if the seawater flowed into international waters.

˅

DURING THIS FINAL PREP PERIOD, Craig was wrapping up the funding for the project. In many ways this was as ambitious and remarkable as the voyage itself. Big-ticket items included technologies like sequencing, which in the early 2000s was astonishingly expensive, and the maintenance budget required to keep a hundred-foot research vessel afloat. Major costs, too, were attached to the need to keep a crew fed, provisioned, and safe on the open sea and in distant ports. JCVI was able to fund some of this, covering the costs of the crew and operation of the vessel. Beyond this, a key funder and supporter was the US Department of Energy's Biological and Environmental Research program, which had been underwriting Craig's projects since the mid-1990s. Key to this support was the program's director, Ari Patrinos, who had provided the first funds to what became the Human Genome Project and who would later launch the DOE's systems biology program called Genome to Life Program, and its Joint Genome Institute. Patrinos also helped initiate the International Panel on Climate Change and the US Global Change Research Program.

Ari Patrinos provided Craig with not only money but also moral and political support, as the DOE paid out substantial grants despite the enmity being directed at him by some in the traditional scientific community. "Craig can be alienating to some people," said Patrinos, "and the reviews were all basically negative about the ocean sampling projects." That mattered, because much of the process of grant-making at DOE and other government agencies is informed by scientific peer review.

"They told me, 'This Venter guy is trying to get you to fund his tour around the world,'" recalled Patrinos. "But I knew what he was doing, combining sailing and research, and I see nothing wrong with having some fun while doing science. So I went against my reviewers, which is something you can do when you're in the Department of Energy—they give you enough rope to hang yourself. You can't do it all the time, but I thought, there are some nuggets in there that could prove very, very useful.'"

Patrinos also came from a background that he said allowed him to understand Craig's ideas about going big with biology. "I started out as an engineer," he said, "and I got my PhD in engineering with applications in things such as atmospheric turbulence and ocean and atmospheric circulations and so on. So I saw the value of different tools as they were developed for advancing bioscience. Being from engineering background also meant that I didn't shy away from big science like Craig was proposing."

"The Office of Science at DOE was dominated by nuclear physicists and high-energy physicists and chemists," said Patrinos, "who were using big tools that advanced their science. Biologists were very, very timid at the time. And very slowly, we pushed hard to push all of this with them. So, for me, Craig was one of these people who were singing my song. I thought there was no reason why the physical scientists should be dominating big science. It was time for biology to go big, which he understood before most others."

Another critical funder was the Gordon and Betty Moore Foundation. It provided several grants, including $24.5 million to procure tools for analyzing *Sorcerer II* data and to make the data available to scientists worldwide. The point person at the Moore Foundation was David Kingsbury, a former assistant director of the National Science Foundation. "I'm a molecular biologist and microbiologist by background," said Kingsbury. "I think I first met Craig when he was a graduate student at UC San Diego. I had just finished there myself. He was a colorful figure in his own unique way, even at that time."

"The foundation's support started in 2003, toward the end of the year," recalled Kingsbury. "We had been looking for an area to support that had great potential, was clearly scientifically ready to develop, but was not being well supported. So internally at the foundation they had decided to support an initiative in marine microbiology, which is when they called me."

"We had this idea that metagenomics would give us an opportunity to start understanding a wide variety of microbes that people didn't know about," he continued. "Microbiology was stuck in a conundrum because they were limited at the time to studying only microbes they could grow, so you didn't even know what was out there. We were interested in Craig because he was able to clone very large fragments of microbial genomes. Then it became clear that you could sequence mixtures of microbes and assemble them individually out of that mixture."

"Given the controversy around Craig, there was a risk, and we knew that," said Kingsbury. "But it was worth the risk to help us better understand the global ecosystem and the role of microbes, their quantity and diversity all over the globe. Later, we had much less community pushback. In fact, we had people coming and saying: 'Look, we need to have a broader geographic representation. Why don't we make sure that Craig goes to wherever and samples there.' So, people started to realize this is a very important approach. And now people use his data routinely."

ON DECEMBER 18, 2003, the great vessel, now fully rigged-out for girding the globe, departed Annapolis. This was a less than ideal date to be heading out into the north Atlantic. Even as the ship motored out of Annapolis, the old city with its domed statehouse and other eighteenth-century buildings was covered with a light dusting of snow. According to the ship's log, the air was crisp and the sky clear. *Sorcerer II* sailed under the sprawling Chesapeake Bay Bridge, past the wintry forests of bare trees covering the low hills on shore, in temperatures hovering near freezing on the crashing gray waves.

On the second day out, December 19, the seas began rising and some fifteen to twenty bottlenose dolphins joined *Sorcerer II*, arcing out of the water and up into the air and back down

again. "We got these really steep waves, and everybody was getting tossed around," remembered Brooke Dill. The crew donned
heavy-weather gear, with Craig and Jeff Hoffman wearing bright
yellow jackets and pants as the ship sailed into a tempest and
the seas got rough. "To be honest, it was some of the worst
weather we sailed in on the world trip," recalled Hoffman.

The storm raged as the ship headed south along the shore-
line off Virginia and the Carolinas, settling down a bit on the
third night out, as *Sorcerer II* approached Charleston, South
Carolina. The wind still wailed and roiled the sea, but by the
time the sun set it was calm enough for Hoffman and his
team to take a sample—"even though," as Hoffman put it, "it
was friggin' cold." The storm's abatement also allowed a film
crew, which had come aboard in Annapolis to shoot a documentary for the Discovery Channel, to set up their equipment in
the stern.

The documentary team was led by producer and director
David Conover. Over the next few months, he and a cameraman
would dip in and out of the expedition in places like the Atlantic
seaboard, Panama, Cocos, and the Galapagos. In 2006, Conover's
Cracking the Ocean Code aired on the Discovery Channel.[6] Gen-
erously, Conover also shared with Craig over a hundred hours of
video and audio that didn't make it into the sixty-minute film—
the source of many quotations in this book.

"So we started down the coast in what started off being
strong winds and ended up being a full-blown gale," Craig told
the camera. He was dressed in heavy, yellow rain gear as the
wind whistled around him. Warnings from the Coast Guard on
the radio were clearly audible in the background. "Your hands

can freeze quite quickly. It's cold out here. Jeff and the team behind me are going to be working for hours filtering these samples in essentially conditions that could support snow, so they are all looking forward to heading south and getting to sampling in the warmer climates."

Craig explained that they were then in waters between the coast and the Gulf Stream—the powerful current that starts in the warm waters of the Gulf of Mexico and travels around Florida then north along the Atlantic seaboard of North America before turning east into the North Atlantic. "This is our cold-water stop near the coast," said Craig. "Tomorrow we'll sample from the Gulf Stream, which will be several degrees warmer."

"The Gulf Stream is like a giant stream of life," he continued. "Every time we go through the Gulf Stream, we have dolphins dancing and fish leaping out of the water, and we'll find out what the bacterial count is. The Sargasso Sea, which is bounded by ocean currents, including the Gulf Stream, is almost devoid of life in comparison to the Gulf Stream. That is why we chose it [the Sargasso Sea], initially thinking we would find low diversity there as our baseline sample. And if that is low diversity, then the whole world is going to be a giant surprise in terms of what we are going to find in this invisible world."

Conover, tall and thin, dressed in orange pants and a thick, olive-green rain jacket with the hood cinched tight around his face, asked Craig, "So why are you taking this expedition around the world?"

"We want to capture as much diversity in the oceans of Earth as possible," answered Craig. "You know, the organisms off Australia are totally different than they are in the Sargasso

Sea, or here off Charleston. We would like to survey the entire world, maybe having as many as ten to twenty billion different genes."

"It's going to help us understand the carbon cycle, [and the] evolution of life on this planet, and maybe give us whole new tools going forward with pharmaceuticals and cancer, changing the way chemistry is done." The expedition might even suggest clues about life on other planets, he added: "We are at the stage where we are thinking of manned missions to Mars and other places to see if there is life there. People will find it bizarre if we find anything, but if we do, what if we find out that it *originated* there?"

∨

AS THE LAST, pale light faded into darkness off the North Carolina coast, Hoffman and his team were hard at work eleven miles from Nag's Head, taking sample number 13. By now, the methods for taking samples had essentially been finalized and would remain the same for the rest of the expedition. The team started the sampling operation by prepping and deploying a blue pump that was Velcro-strapped onto a long pole and dropped into the sea. Usually, the pump descended about eight feet into the water, although the depth varied a bit. The long pole was a spinnaker pole modified by Cyrus Foote, which Brooke Dill remembers as being "like a boom arm, that we could drop down to the right depth and take all the readings and filter forty to a hundred liters of water. (Differences in volume usually had to do with turbidity, or the level of mud, plants, and other particulates in the water.) A "multiparameter" instrument called the YSI 6600 was also strapped to the pole to simultaneously measure salinity, temperature, pH, dissolved oxygen, and depth. All this equipment was sterilized to avoid contamination by the humans handling the samples.

The samples were then filtered through the three progressively finer filters the scientists had been using since Halifax, each round sheet the size of a medium vinyl record disc. Once the water was pumped through them, the filters were sealed with buffers and stored in the onboard refrigerator at minus-80 degrees Celsius, ready to be shipped back to JCVI on dry ice at the next port with an airport.[7]

Brooke Dill had this to say about collecting the samples: "What I loved about it was that it was really simple, and kind of cathartic. What we did was to take water and push it through the filters using compressors." The compressors were housed in the engine room. "You had to monitor what was coming in, to make sure there wasn't too much stuff in the water, so that we didn't blow anything up and the bilges didn't get too clogged. There was a sensor that came off the boom that would tell us the pump was pulling up water. At the same time, we would record temperature, pH, and the rest, and then we would write down the latitude-longitude to make sure we had the exact location where we were. And then, after the filters were done, they were vacuum-packed."

"Most of the time we'd rock out to music while we were working," she added. "Jeff liked heavy metal but played classic rock when I was back there."

Later, scientists at Craig's institute would analyze the samples taken from Halifax to Florida, and beyond. They discovered a wide variety of microbial populations in individual samples, depending on where they were scooped up from the sea. Depth played a big role in yielding these differences. Variations in temperature and salinity also mattered. In a 2007 paper in *PLoS Biology*, JCVI scientists reported that samples taken during the early days of the expedition off the eastern US ended up falling into "two well-defined clusters."[8] The five samples taken along the route from Nova Scotia through the Gulf of Maine contained similar communities of bacteria in deep seas and shallower, cold water. Further south, samples taken in the estuaries of Chesapeake Bay and Delaware Bay were very similar to one another, and markedly different from the northern samples. "Estuaries are complex hydrodynamic environments that exhibit strong gradients in oxygen, nutrients, organic matter, and salinity and are heavily impacted by anthropogenic nutrients," noted the 2007 *PLoS Biology* paper, by first author Douglas Rusch, a JCVI computational biologist, and a team of coauthors.[9] Yet, "the Bay of Fundy estuary sample . . . clearly did not group with the two other estuaries, but rather with the northern subgroup, perhaps reflecting differences in the rate or degree of mixing [of salt water and fresh water] at the sampling site." The genomics of the bacteria collected also showed differences in how organisms react to and live in these various locales.

THE NEXT AFTERNOON, on December 21, 2003, after another long day of filming and sampling, *Sorcerer II* reached the sea off Charleston, South Carolina. As the weather turned rough again, the howling winds and drenching rain sent the exhausted scientists, filmmakers, and crew below to ride it out—not realizing that they were facing potential disaster and one of the most serious challenges of the entire voyage.

Outside, the winds blasted and rain, mixed with hail, pelted the main cabin in a steady rat-a-tat-tat. Charlie Howard ordered the first mate to check and make sure the anchor, which was winched up in the bow, was secure. With near gale-force winds blowing, Howard was concerned it might come loose and fall against the hull of the boat, which could easily smash a hole in its side. Normally, Charlie would have done this himself, but he had just had knee surgery. So he dispatched Brooke Dill and another crew member we'll call Kim—she ended up getting fired, so we'll change her name here—to check it out.

Dill later recalled the stormy scene: "We went up to the bow and we had waves crashing over us so much that [Kim's] life jacket inflated." But "the sun was setting, and we had to lash down the anchor." As Kim started tying knots, "I was watching her, and I wasn't sure she was getting it done—you know, the knots—and I asked if she was sure that she had everything. And she said, 'yeah,' and ordered me to head back to the cabin. I questioned her twice more, and she yelled at me, saying everything was okay."

Soon after, with both back in the cabin, the crew began to hear strange noises coming from the bow of the ship, a banging and crunching that rose even above the cacophony of sounds from the thrashing storm. "We all kept saying," recalled Heather Kowalski, "do you hear that noise? I think I hear a noise." Charlie Howard asked Kim if she was sure the anchor was secured. Kim was adamant that it was.

The noises continued as most of the crew hit their bunks and tried to sleep in the roiling waves. "It got so bad that in the middle of the night I woke up, hearing all these noises because

I was sleeping up forward," remembered Hoffman. He got up and found Charlie and Kim and most of the crew and passengers up, too, in the main cabin. "Charlie was asking the first mate again if she stowed the anchor," said Hoffman, "and she said, 'yeah, yeah, yeah.' She was defiant. So, Charlie sent me and Cyrus up there—and you've got to imagine now, I didn't know anything about the boat at the time. We could see we were getting bashed with waves, and it was cold and rainy. And yeah, the anchor was out. There was like twenty feet of chain that was just under the boat, bashing into the hull."

"The anchor being undone, it was plowing into the fiberglass hull of the boat," remembered Kowalski. "I mean it could've easily put a hole in the side and sunk the boat."

"Apparently, there had been a twist in the chain," said Charlie Howard, "and the mate, who was not particularly experienced with the anchor, had not made it completely secure. And there was no safety on the chain if the lock happened to pop off—and pop off it did in that heavy weather. As the bow dropped between the rolling waves there was a slam and crash under the bow of the boat. Very scary! Fortunately, there wasn't any serious damage, but there could have been."

By the next day the storm had lessened considerably, and then the weather miraculously cleared into brilliant blue skies as *Sorcerer II* reached Fort Lauderdale. It was a relief to see the glimmering white high-rises arrayed in the late afternoon light along equally white beaches. They would spend the next two weeks in this city repairing the boat and preparing for the next leg, around the Florida peninsula and then across the Gulf of Mexico and the Caribbean to Panama. And it was here that, for the first time on the voyage, Craig showed up sporting a beard.

For years, he had tended to be closely shorn, but the whiskers softened his face a bit. Since then, it's a look he has seldom been without.

⩔

AFTER THE FRIGID ordeal off the Carolinas, the crew reveled in the warmth and comfort of Fort Lauderdale. They celebrated Christmas and held a New Year's Eve bash, all of them decked out in *Sorcerer II* hats and T-shirts. Departing the United States, finally, on January 7, the ship was accompanied once again by dolphins leaping so gracefully they looked like they were dancing, *Sorcerer II* set sail, slicing through gentle, phosphorescent blue waves as they headed south. An hour later they passed the slick towers of Miami. *Sorcerer II* kept going, picking up speed.

The crew now included Craig, Charlie Howard, Jeff Hoffman, Cyrus Foote, Brooke Dill, Kathy Urpani as cook, and Tyler Osgood, one of the institute's information technology experts. Guests on board included Venter Institute trustee Dave Kiernan and a sailmaker and friend of Charlie Howard named Scott Gibbs. (Conover's film crew had debarked in Ft. Lauderdale and would rejoin the voyage in Panama.) That afternoon, *Sorcerer II* reached the upper Florida Keys, a pearl string of islets extending from the southeast corner of Florida's mainland peninsula westward into the Gulf. First in this archipelago is Little Torch Key, south of Miami, followed by dozens of lush green islands set off by white sand against the turquois sea, extending to Key West and winding up with the Dry Tortugas at the tip.

As stunning as this progression of gemstone-like islands was, the route was torturous. Charlie Howard labored to navigate a narrow channel between, to their west, a wall of colorful but jagged coral reefs just under the surface and a wall of water to their east, where the Gulf Stream rushes by the islands at three knots heading north—in the opposite direction they were headed. "There isn't much space between the two," said Howard. "We were squeezed into a tight corridor between reefs and the Gulf Stream."

Adding to the excitement, recalled Howard, the wind had increased to gale force, blowing at twenty-five to thirty knots from the gulf coast of Mississippi, Louisiana, and Texas. The islands between the ship and the Gulf of Mexico blunted these winds somewhat, but the waves were still eight feet high. "I was on watch the third night out," said Hoffman, "and the winds were bringing us large waves running from the east. This created a very confused and uncomfortable sea." After his watch, "sleeping was almost impossible as we thrashed from wave to wave."

The ocean remained mostly rough with high, bucking waves for the next five days as *Sorcerer II* sailed across the Gulf of Mexico and the western Caribbean toward the isthmus of Panama. The gyrating sea and high winds took their toll in various ways. At one point, when a huge wave smacked the ship sideways, a spare computer flew out of a cabinet and smashed. Cook Kathy Urpani had to deal with an oven on the fritz. The crew, however, managed to settle into a steady flow of watches, meals, and maintenance, the throttling up and down of engines and the rising and falling of sails. When they weren't on duty, they napped, read, wrote, or simply did nothing. In the evenings, Urpani entertained everyone with her guitar.

Sailing past Cuba, the crew that was awake in the wee hours saw fires burning on land as *Sorcerer II* turned south around the western end of the island and through the Yucatán Channel between Cuba and Mexico. The sun rose that morning over Cuba as expansive, golden rays emanating from gray clouds. "We can smell Cuba with the warm southeastern breeze blowing across the land and out to *Sorcerer II*," wrote Craig in the expedition's blog. "It smells organic, like Mexico or India, a mixture of humanity, dry land, diesel exhaust and agriculture."

On January 10, Craig wrote in the ship's log that he came on watch at 5:00 AM. "As soon as I came on deck, I saw the Southern Cross for the first time on the expedition," he wrote. "It was low on the southern horizon. The nearly full moon was behind and overhead."

The expedition took only a few samples on this leg because some of the countries along the way refused permission. There are essentially no international waters in the Caribbean Sea, so the team was required to obtain permits from every government along the route. Nicaragua denied permission, but Honduras said yes. When they did manage to sample Gulf and Caribbean water, they found that the temperature difference between the samples was a full 10 degrees Fahrenheit, from 81 degrees in the Gulf Stream to 71 degrees just outside it. The warmer water, said Hoffman, filtered faster due to its clearness. The cooler water was less clear, with more microbial life coalescing like scum on the paper filters in splotches of greens, yellows, and oranges.

⌄

ON JANUARY 12, the coast of Panama loomed up from the sea as a dark bank of low hills covered with lush, olive-green jungle. *Sorcerer II* motored into a half-moon-shaped bay off the small city of Cristobal in the district of Colon, and waited for its turn to enter the Panama Canal. To celebrate the crossing, the crew broke out Corona beer, Jack Daniels, and tequila.

"We had to wait a day or two on the east side for our turn to go through the canal," said Hoffman. "We went into town. It was really sketchy. We went to this market. It was all fenced in with armed guards." The team was told to be on guard for thieves in a town known to be somewhat lawless. "Everyone warned us at the marina to be careful."

The canal that *Sorcerer II* was entering cuts through the Isthmus of Panama from shore to shore, a span of about forty miles. It was finished in 1914, only after thousands of workers, mostly people of color from the Caribbean, had died during its construction from disease and heat. From its entrance on the Atlantic side in Limón Bay, west of Colón, this transection of the American landmass first runs due south to the Gatún Locks—three long, slender cavities carved into the ground that use gravity to shunt water in and out of the inland Gatun Lake. Two more sets of locks on the western side of the isthmus do the same thing using water from the lakes of Alajuela and Miraflores. The locks, all identical in size, were built in pairs to allow for simultaneous transit of ships in both directions.

On the eastern side, as ships pass one by one into the locks, the cavities fill up with water to lift them ultimately eighty-five feet above sea level. This is the altitude of Gatun Lake, an artificial body of water built when the canal was being created and fed by waters from rain forest streams. The long, narrow lake splays across the center of the isthmus in a southwest-to-northeast direction, and is the open body of water ships sail through to get from the eastern locks to the western locks. On the Pacific side, the locks then lower the ships back down to sea level in the Bay of Panama. And vice versa for those sailing eastward.

Early the next morning, at 5:30 AM, Javier, the pilot assigned to guide *Sorcerer II* through the canal, arrived and announced that the ship had to leave right away for the locks. Its turn had come. "We raised the anchor and headed for the first lock," said Craig, "taking us out from the Caribbean Sea on our way to the Pacific."

About a third of the way across the isthmus, *Sorcerer II* stopped by Barro Colorado Island in Gatun Lake, home to the Smithsonian Tropical Research Institute. As *Sorcerer II* approached, the complex appeared from the water: a cluster of low buildings rising up a hill on this jungle-clad island where around four hundred resident researchers at any given time are studying the rain forest in central Panama, and over fourteen hundred researchers visit each year. On January 16, Craig gave a lecture there. Wearing a green Hawaiian shirt and khakis, he showed PowerPoint slides of the trip so far, talked about the data published in the Sargasso Sea paper, and shared his expectations of what might ultimately come out of the global voyage.

On Barro Colorado, the team also met Eldredge "Biff" Bermingham, a geneticist who was then chief scientist—later he was appointed director—of the institute. Bermingham and his team had worked with Craig and others at JCVI to set up this visit. That afternoon, he came on board *Sorcerer II* to be interviewed by the documentary film crew and expressed enthusiasm for the effort: "I don't think anyone, having seen what Craig has discovered in a very small quantity of water off the island of Bermuda, would question if there are going to be applications for the genetic diversity that he is uncovering."

Bermingham also joined Craig and Jeff Hoffman to take a sample out of Gatun Lake, with the film crew capturing a moment when Bermingham and Craig checked readings from instruments measuring the condition of the water.

"The salinity is .06, pretty fresh, we won't get much salt out of that," notes Craig on the video, reading off a handheld display. "The temperature is 28.5 degrees [Celsius]. That's warm. The pH is stabilizing a little bit, but it's still 7.8, 7.79, so it doesn't look too bad."

Eldredge "Biff" Bermingham (center) with Craig Venter (left), taking soil samples in Barro Colorado Island, Panama, January 2004.

Bermingham is interested in the process: "How long does it take to calibrate?"

Hoffman chimes in now, leaning over as he works the pump. "About a minute. I let it flush out water for about two minutes, give it time to clean out the hoses."

Most visitors to Barro Colorado take the opportunity to hike its extensive trails through the jungle. For the *Sorcerer II* crew this was a chance to take the first soil sample of the expedition—a procedure they would repeat in other exotic spots to compare with the microbes they were finding in oceans, lakes, and streams. With Biff Bermingham as guide, the group walked under the dense canopy of the rain forest, encountering howler monkeys and orchids thriving against ceiba trees so huge they've grown giant buttresses resembling moss-covered rocket fins. Later, Brooke Dill related a moment from the hike: "I remember Craig going underneath where the monkeys were and they were shaking the branch at him. And I said to Craig: 'don't go underneath them, they'll throw their shit at you.' And he laughed at me. And they did."

Leaving Barro Colorado, *Sorcerer II* continued across Gatun Lake to the Pedro Miguel Locks. They brought the ship down to Miraflores Lake's elevation of fifty-two feet and finally, through a last set of locks, down to sea level. Remaining to be traversed was only a seven-mile waterway dug out of the isthmus, and *Sorcerer II* entered the water of the Pacific Ocean, sailing into the wide, sweeping Panama Bay. The office towers and hotels of Panama City glittered along its curving white beaches as the ship proceeded toward the tip of a string of causeway islands jutting into the bay. Choosing a spot near the Fuerte Amador Resort and Marina, the crew dropped anchor.

On Barro Colorado, Bermingham had described on camera how the skinny isthmus of Panama separated two huge ocean ecosystems, each with its unique communities of microbes. In particular, he said, "on the Pacific side, the temperature swings are much more striking than what you get in the Caribbean. It can get up around twenty-six to twenty-seven degrees [Celsius] during the summer months, which is during the rainy season. During the dry season, when the wind is pushing across and changing the current regime, the water gets much colder, down to around twelve, thirteen, fourteen degrees. You get this tremendous upwelling of nutrients in the Bay of Panama. You go a little bit farther west, and you lose a little bit of that upwelling close to shore. But you get it again as you move offshore. And that's going to have a tremendous effect on the microbial community."

"So, you have the marine diversity that you find in the Caribbean and the marine diversity that you find in the eastern Pacific," continued Bermingham, "and you also get this extraordinary diversity that you find sitting in between, on the isthmus of Panama, in the lakes and canals. You see that diversity because the elevation changes and the rainfall gradient changes, and all that translates into the difference of species richness—much of which we still don't know much about."

"I think in evolutionary biology, which is what a lot of us are interested in, there's the hope that we'll be able to use some of the approaches used by Craig to better understand how species proliferate both over space and over time," he said. "But we're not there yet. Every time we apply genetic approaches to the study of either flora or fauna in the tropics, we always find that there is more diversity beneath the surface than was anticipated. So we're always discovering more species. Of course, when we get to microbes, we know nothing about them. Zero."

Y

ON JANUARY 19, *Sorcerer II* left Panama City and the Americas. It was a bright day with high, rolling puffs of white clouds when the ship turned its bow west and left the sight of land for the open sea, headed to Cocos Island, about 580 miles away. The team stopped about thirty-five miles out in calm seas to take sample 21. Raising the sails again, they were happy to make excellent time—but shocked to sight something they had heard about but hadn't realized was, well, *everywhere.*

Plastic. Lots and lots of plastic.

They were approaching a region in the eastern Pacific where multiple currents converge into a huge gyre that traps an endless stream of plastic—from toys, plates, flip-flops and sunglasses to chunks of hull and deck from boats, and on and on. "One day," recalled Hoffman, "a refrigerator floated by."

"Styrofoam was always the main ingredient in a soup of anything from oil drums to plastic bottles," reported Kathy Urpani in the ship's log, "a smorgasbord of plastics which we plucked out of the ocean on a daily basis, liberating little triggerfish and crab lodgers as we filled maxi sized contractor bags of rubbish."*

---

* *Sorcerer II*'s logs during this phase of the project are from unpublished copies kept in the JCVI archives. Entries from the *Sorcerer* logs have been edited slightly for length and clarity.

In the years since the global expedition, scientists at JCVI have been collecting detritus like this from the ocean to study the microbial communities living on and in the plastic. According to JCVI oceanographer Chris Dupont, who spearheads these studies, some of the species found on the trash contain genetic variants of terrestrial bacteria that are not usually seen in the ocean. "They come from people handling the plastic on land, and from other land-based microbes that cling to the plastic as it makes its way down freshwater streams and into the sea," he said.

The crew took two more samples on the way to Cocos—one at two hundred fifty miles and the other at thirty miles out from the island. "The winds were light most of the way," said Craig, "and we had to do a great deal of motoring. However, we had frequent squalls with dramatic effects, like increased winds and spectacular skylines and sometimes rainbows. The moon was new, so the night sky was awesome. Toward starboard right, we had the Big Dipper pointing to the North Star, and to our port we had the Southern Cross. The Milky Way was the brightest that I have ever seen, and the moons of Jupiter were visible with the naked eye."

On January 21, they first caught sight of Cocos, a magical island off the coast of Costa Rica where some scenes of the *Jurassic Park* movies were shot. Cocos is mostly uninhabited and is inundated with rain year-round. Across the centuries, this made it a favorite stop for ships looking to load up with fresh water before heading out into the wide-open portions of the Pacific. "When you read about Darwin and the *Beagle*," explained Craig to the film crew, "what was the biggest problem in crossing the Pacific? It was the lack of fresh water. It rains here all the time. You just go right into this anchorage, and there are three or four waterfalls falling right into the ocean. Boats used to sail right under them to fill up barrels on their decks."

Giant boulders and rock faces near the falls are covered with the names of visiting sailors and ships dating back hundreds of years. "There is a big one from Jacques Cousteau's boat in 1997," said Hoffman. "An engravement there on a stone."

On the second day anchored at the island, Craig and Heather Kowalski took a hike into the interior, reveling in the lush, dense jungle fed by all that water. A less welcome discovery was a colony of fire ants that attacked them during a rest stop, stinging them and covering them with welts.

On January 31, *Sorcerer II* departed Cocos Island for the Galapagos. The crew saw a distant pod of orcas and a spot where the water looked like it was "boiling" with a thick cluster of thrashing fish. Brooke Dill, writing in the log, called this and other similar phenomena "fish balls"—"a teeming ball of fish scared together by some unseen predators. At times we can grab a glimpse of the unknown as monster tuna, sharks and other toothed fishes push their dinner to the surface, trapping them helplessly."

One evening at midnight the crew celebrated crossing the equator. "We are told to write poetry, prepare a gift and dress as a creature of the sea the following evening," reported the ship's blog. "At the designated time, organisms of the deep appeared on deck under our white party lights, and foods are scattered upon the table to begin the feast. . . . It is customary for those onboard who have crossed the equator by boat to lead this type of celebration and passage for the others who have never crossed before. . . . Captain Charlie leads the celebration

On Cocos Island, Costa Rica, a fresh waterfall onto the beach. Historically, these waterfalls were a major source of drinking water for sailing ships.

When sailing ships stopped for water in Cocos Island, Costa Rica, they often carved names into the boulders on the beach. Clearly visible is the carved date of March 2, 1839.

dressed as Poseidon ... [and] after offerings, trials, and words said, a ritual grog was passed, and we all sat together in glory under the starry lit night in this black sea to a feast celebrating the sea."

AFTER A NINE-DAY JOURNEY, the longest time yet at sea for the expedition, *Sorcerer II* arrived in the Galapagos Islands on February 2, a day everyone later remembered as extremely hot. The expedition would stay in these islands for almost six weeks, taking samples not only in the ocean, but also in salt ponds, inland lakes, sulfur vents, and mangrove swamps.

Sorcerer II anchored in a bay on Cocos Island, January 2004, along with the Discovery Channel dive support ship.

"We awoke to the stark beauty of the Galapagos," logged Brooke Dill the morning they arrived. From the sea, the low volcanic islands looked like the dark shells of the giant tortoises that live in the Galapagos. "We have stepped back in time to a primitive land," she continued. "This is a place where the very instinct to survive is in the air and the competition is fierce. Our first Galapagonians were very dark and large bottlenose dolphins that charged to our bow. Boobies and frigates soar above us as we find flocks of dark-rumped petrels floating together in the slicks, looking like tiny cotton balls with dark tops swept along in the current line. As we approach, they take flight in a burst of energy."

The official name of the Galapagos, part of Ecuador, is the Archipiélago de Colón. In Spanish they're also known as Las Islas Galápagos, having been originally discovered by chance in 1535 by a Spanish priest living in Panama named Fray Tomás de Berlanga. The priest was the first to describe the strange creatures he saw there. Flemish cartographer Abraham Ortelius called the island chain *Insulae de los Galopegos*—the islands of the tortoises—on a world map he published in 1570.

In 1684, an English pirate named Ambrose Cowley arrived and gave some of the islands anglicized names, as did other English visitors. This is why many of the islands have more than one name. Espanola is also called Hood Island, after a British Lord; Santiago is James Island, after a member of Cowley's pirate crew; Santa Cruz is also called Indefatigable Island, after a British warship. The islands remained largely untouched for centuries, including when Darwin visited them in the 1830s. During the twentieth century more people came, including tourists and researchers. They overhunted many of the bizarre animals, including the giant tortoises, reducing populations.

Visitors over the years also introduced outside species such as goats to provide a ready source of meat, and blackberries to grow as a crop. These invaders and others quickly proliferated, squeezing out local fauna and flora. In recent years, the Galapagos government has enacted strict controls on bringing in any non-native macro species. Micro species, however, are another matter. Microbes still come in on boats and planes, in luggage, and on people. New microspecies also have arrived naturally for millions of years, carried by winds, birds, and ocean currents. The impact of these micro-invaders is largely unknown.

Sorcerer II dropped anchor off Santa Cruz Island at Puerto Ayora, the largest town in the Galapagos, with twelve thousand people. This became the home base of the expedition over the next few weeks as the team explored the islands of this archipelago some seven hundred miles off the coast of South America. Craig rented a house on a hill above the main street, the Avenida Charles Darwin, which begins at the town's main dock. This thoroughfare finishes at the other end of town at the Charles Darwin Research Station, a complex of labs and offices, a library, and an exhibition space opened in 1959 by the United Nations.[10] The station's Ecuadorian and international staff helped Bob Friedman and Karla Heidelberg make arrangements and secure permits for the team's visit, although it later turned out that some of the station's staff weren't too keen on *Sorcerer II* being there.

"I remember the harbor at Puerto Ayora was kind of tight," said Hoffman. "They don't normally get huge hundred-foot boats there, there's a lot of fishing boats and dive boats. So getting into where we were going to anchor was nerve-racking." The town itself was filled with waterfront artisans' shops, restaurants, internet pubs, and hotels. "The island smells of volcanic dust and wind," wrote Brooke Dill, "dry wood and an occasional fish. We can easily walk the distance of this village on its cobblestone streets in 10 minutes. Sea lions bask on the waterfront, brown boobies and pelicans plunge into the water, marine iguanas lie lazily on the rocks, flocks of golden cow nose rays can be seen gliding across the bay, while water taxis busily ferry people to and from the village."

Science writer James Shreeve dropped in on the expedition a few weeks later to gather information for an article for *Wired*. The piece he would ultimately publish describes the aims and planning of the Galapagos sojourn, and some of the team's expectations about what they would find. It also noted the inspiration Craig took from Charles Darwin:

> Venter conceived of his expedition as following in Darwin's footsteps, and now he was sailing into the same bays and trudging up the same rocky paths as the great man himself. He had organized the visit well ahead with scientists at the Charles Darwin Research Station on the island of Santa Cruz to ensure sampling in the most productive spots; these included several unusual environments, each likely to contain a unique spectrum of microbial life, with differing metabolic pathways and hence different sets of genes.[11]

The expedition's first excursion in the Galapagos was on land, venturing into the interior of Santa Cruz Island, where historian and local guide Greg Estes took them to a section pocked with "misty, primordial-looking craters," as Dill would describe them in the log. From there they moved on to a forest full of giant daisies. Darwin collected these in the 1830s, discovering that these plants appeared on several islands and yet were different. "It was his largest representation of species radiation," wrote Dill. The group also saw frigate birds taking flight after bathing in small inland lakes, their seven-foot wingspans in this weird landscape giving them the look of pterodactyls from millions of years ago.

Rounding out the crew during the Galapagos explorations were a park ranger named Simone and oceanographer and microbiologist Victor Gallardo, a professor in the oceanography department at the University of Concepción in southern Chile. Craig had met Gallardo during a visit to Chile and learned about his expertise in the ocean between South America and the Galapagos. Gallardo was at that time in the midst of a ten-year study of the nutrients that feed ocean life in the southeastern Pacific off Chile, including the microscopic organisms at the bottom of the food chain that feed the region's phytoplankton and other microbes. Later Gallardo would call his stint with *Sorcerer II* in the Galapagos one of the most exciting projects he ever did, and a source of ongoing learning: "I'm still working on understanding some of the discoveries we made in the Galapagos."

One unwelcome development in the Galapagos was the resurgence of claims by some critics that Craig was secretly out to steal the micro-patrimony of the islands. JCVI bio-diplomats had worked for weeks on an agreement with the Ecuadorian government in its embassy in Washington, DC, and with officials in the Ecuadorian capital of Quito. Despite these efforts, on March 10, 2004, the ETC Group—the same organization that called Craig a "biopirate" in Bermuda—published a "communique" on its website titled: "Playing God in the Galapagos: J. Craig Venter, Master and Commander of Genomics, on Global Expedition to Collect Microbial Diversity for Engineering Life."[12] Again accusing Craig of exploitation, the group noted that one of his institutes was being funded by the Department of Energy "to create new life forms in the laboratory that could be engineered to produce energy or clean up greenhouse gases. Exotic microbes—such as those found in the Galapagos—are the raw materials for creating new energy sources and new life forms."

Bob Friedman, Greg Estes, Heather Kowalski, and Craig Venter on Daphne Major, Galapagos, Ecuador, February 2004.

The negative publicity led to a peculiar situation whereby the expedition, having been given permission to take samples, was then forbidden to remove them from the islands. "They said we would be arrested if we tried," recalled Craig. "They were afraid that we were going to steal this stuff and make millions." Bob Friedman remembers that officials in Ecuador and in the Galapagos had issued permits and sent emails giving the okay to fly the samples back to the States. "Then something strange happened," he said. "Apparently, some academic lawyer in Sweden"—yes, Sweden— "heard about the expedition and started to talk to the Ecuadorian ambassador in Stockholm. Then this ambassador contacted somebody in Ecuador's Ministry of Parks, or the Ministry of Environment, it wasn't clear which one, and basically told them that the microbes in the samples were 'the diamonds of the sea,' and convinced them that Craig was up to something."

"Then someone took this to the Charles Darwin Station," continued Friedman, "and said, 'well, you guys are biologists. Could you figure this out?' And it was tasked to the head of the Marine Conservation Group, a German marine biologist at the station who stepped in to stir things up." Curiously, said Friedman, the local Galapagoan authorities on the islands were mostly ambivalent about what the gringos on *Sorcerer II* were doing. "It was the non-Ecuadorians at the Darwin Institute that were complaining."

It got even weirder as a Swedish lawyer in Stockholm—a friend of the original academic—joined the ongoing negotiations with the Ecuadorian authorities in Quito and in the Galapagos. "He insisted that we needed a Memorandum of Understanding, additional permissions," said Friedman. "This, coming from an academic in Sweden who was determined to protect the genetic patrimony of Ecuador," whether the Ecuadorians wanted this or not.

As politics broiled and negotiations continued, Craig and the team were still allowed to take samples, which they started collecting on February 4. They sailed south from Puerto Ayora to Isla Floreana, where they brought on board samples 27 and 28, and then headed to Santiago Island to take sample 29. On February 9, the crew awoke at 3:00 AM to head to possibly the most unusual site on the entire voyage, an underwater thermal vent next to a small, remnant volcano island called Roca Redonda. There, fifty-seven feet below the surface, the vent exuded a hot froth of heated water and hydrogen sulfide, bubbling up from an underground bed of lava below the ocean floor. It was a crazy place for life, even microscopic life, to exist.

According to Brooke Dill, the team planned to take a sample directly from the vent rather than using the pump on the boat. "The conditions are very rough at this site," she logged as the group headed to Roca Redonda in the ship's dinghy. "There are two to three knot currents, cresting waves at the anchor spot. Worse conditions exist at the proposed dive site, but this is an opportunity at such an unusual site as sulfur bubbles come up from the vents and the water becomes a balmy 82 degrees Fahrenheit."

"My personal job is professional dive buddy to Craig," continued Dill at the site itself. "Geared up and ready to go, I watch as Craig flips backwards off the dinghy with a Galapagos dive guide and they fight their way in a ripping current and crashing waves to a line that will take us to the vents. I follow directly behind and drop to about 40 feet when the sea floor comes into view. I motion to Craig that I will bring down the YSI probe [a sensor to gauge ocean conditions] and tubing, then watch as they hand-over-hand drag themselves against the current and downward." Right behind Dill came the documentary film crew, who had their own issues conveying their camera and equipment into this tumbling cauldron of currents.

"After fighting the surface conditions once again, with the instruments in hand," wrote Dill, "I pull my way back down to the film crew and Craig who have chosen the site for sampling. I hand it over and watch as Craig claws his way to the vent, pulling the tubing as hard as he can against the currents. He is on his hands and knees with a determined look on his face. I can tell that he no longer notices the camera in his face and the lighting but sees exactly what he wants and seems quite determined to get it. With currents shifting and sea lions dipping over our heads, he never looks up but fights to hold the instruments in place."

"It was quite a dive going down there," said Craig afterward to the video crew, "but experiencing that biology and putting my hand in the warm vent and taking the samples directly, it was a real connection to the biology. I can't wait to get those samples back to the laboratory."

Craig had targeted this seep hoping to find microbes that would hold clues to how they produce hydrogen. Like oxygen and other chemicals, hydrogen is a by-product of certain bacterial species' natural processes. "One way to look at bacteria," said Victor Gallardo, "is that they are like little factories that use their cell machinery to function and to handle inputs and produce outputs like hydrogen."

While the team at Roca Redonda was in the water, Gallardo had a near brush with death. He had stayed on board *Sorcerer II*, asking one of the diver-guides to collect a plant specimen from the bottom in a glass bottle. "I wanted to check it for microbes," he recalled years later, planning to use the high-powered microscope in the ship's lab. "This plant was collected from where the gas was seeping, although we didn't yet know what kind of gas," said Gallardo. "So this diver, he came up and left the bottle for me and then left to do some more diving. I was on the yacht, practically alone. Maybe someone was on deck."

Gallardo remembered thinking to himself, "I'm going to open this thing and find out what kind of grass is this." But when he opened it up, he must have had the bottle too close to his nose: "It was sulfide gas, which is very deadly. It really hit me, so actually I set it down and I took some breaths. I didn't know what to do, I just waited to see what would happen, because people are diving, and there's only one deck man there, so I wonder if I'm going to die." Gallardo figured he must have passed out. "About five hours later, people had returned to the boat, and they saw me waking up. And they said, 'Wow, what a siesta you took, five hours of siesta.'"

Later that afternoon, after Gallardo had recovered and the divers and equipment were back aboard the ship, he took a different bottle of water from the vent—removing the lid with more care this time—and peeked at some droplets through the microscope in the research area of the main cabin.

That evening Craig, Karla Heidelberg, and others crowded around the Chilean scientist to check out the tiny microbes. On a monitor carrying a live feed of the images Gallardo was seeing through the microscope; they watched a progression of bizarre-looking microbes with spikes and rods that looked like sheer wings. Describing the scene that would later air on the Discovery Channel, Dill wrote of the "diatoms, bacteria, and ciliates" on the screen and the reaction of her colleagues: "They are pointing and excitedly exclaiming, 'oh!' and 'look here!'

"So, check out this guy here," said Craig, pointing at a huge microbe that looked like a cylinder with a fright wig on one of its ends. Other organisms drifted and motored across the scene, microbes that were spherical, fuzzy, long bodied, and more.

"Some of these ones are confused," said Gallardo.

Craig peered at another organism. "What about this guy right here?"

"That one could be a cyanobacteria, the very slender kind," said Gallardo, pointing at a long, cylindrical creature, "but it would require more fine analysis." Cyanobacteria are a critical community of photosynthetic bacteria that inhabit the surface waters of every ocean and sea around the world and produce a significant percentage of the oxygen in Earth's atmosphere. "These are remnants of the leaves," he said of some floating particles. "Plant material."

"Here comes the elephant back through," said Craig as the huge cylindrical bug floated again into view.

"The rotting of this material causes sulfides, and that is the food for these guys," said Gallardo. "These bacteria are constantly cleaning up the whole system that allows this ecosystem to live. From bacteria to penguins to all the birds that we saw today, and the turtles, the whole thing is tied up in a very finely working system."

"Look at this guy just spinning in circles," added Karla Heidelberg. "It's like a microbial circus."

Years later, Gallardo remembered seeing something that surprised him that afternoon: "a filamentary bacteria." He had never seen one before, despite studying bacteria from sulfur vents for years. "These are bacteria without pigmentation," he explained, "because otherwise, they would have been cyano-bacteria." The chlorophyll in cyanobacteria gives it a green color. "But they were not cyanobacteria, they were multicellular. And they were mobile. Some of them were really lively."

"I was looking for a very large bacteria called *Thioploca*," he recalled, referring to a genus known to have a vacuole (chamber) containing sulfide and nitrate. It makes energy by oxidizing the sulfide with the nitrate. "This was a discovery that I made in 1963, I think it was. But these guys, they were totally different. And the first time that I looked at them in the microscope, there were many of them. Many of these bacteria were slenderer than the previous ones that we knew. I later published a paper in Spain with my wife, and I divided this type of bacteria into two groups: the ones that we called mega-bacteria, which are the ones that I found earlier, off northern Chile, and another one which we called microbacteria, which are much smaller and, much more diverse."

"This was so different," said Gallardo, "I couldn't believe that somebody had not studied them or hadn't found them before. So I looked for the literature everywhere, and at most there were some hints. People finding something that they called slimy grass, because that's what it looked like, at a depth within the continental shelf, no deeper than that. Then nothing. So I said, 'Well, maybe this has been found in fossils.' And then I started looking at the paleontological literature and I found loads of information about fossils. And my good friend Bill Schopf," a geobiologist at UCLA, "he received my photos and my description, and he said: 'Wow, Victor!' He was very, very impressed. Very happy, also, since many of those forms he had seen as fossils in rocks as old as 3.4 billion years old."

"And they had always thought they were early cyanobacteria," said Gallardo. "And so, if they had been cyanobacteria, then maybe there was oxygen on Earth really early. But then I told him: 'These guys don't need oxygen. They live in the sulfide.' So, we started communicating and publishing together. And then it took another turn, because Bill Schopf was attached to NASA's program looking for fossils in rocks being collecting on Mars. Now they're looking for fossils like these on Mars, an unexpected outcome of a simple finding that day in the Galapagos on *Sorcerer II*."

Two days after their work at the Roca Redonda seep, the team sailed to Isabella Island, where they set out to take another unusual sample—this time in a mangrove swamp, where clusters of tall trees and their tangled roots were broken up by pools of brackish water tinted brown-orange by the mud. "Juvenile fish safely maneuver through the roots of the trees," wrote Brooke Dill. "Sea turtles glide below the glassy surface of the water and bury in the mucky mud to rest, with some of them emerging from their hideouts in huge clouds of muck as the Zodiac carrying the scientists passed by."

In the documentary that ultimately aired on Discovery Channel, the Zodiac stops and a shirtless and barefoot Craig leaps over the side into the red, waist-deep water. "I'm knee-deep in muck," he said. "The number of little guys living in this have to be prehistoric billions. I don't know if you can smell it: just sulfur pouring out. Sulfur [is produced by a] primitive metabolism, so there have to be a lot of cellulases that break down to plant debris. It's primordial ooze."

"It's an unbelievable mixture of fresh and saltwater, it mixes right in this little pool at the entrance," continued Craig, who found himself standing near a dozen or so giant turtles swimming and seemingly playing, another scene that felt like Earth millions of years ago. As for microbes, Craig said on the video that he was expecting to find species that thrive in a mixture of salt and freshwater, and also in sulfur. As he spoke, he collected a sample of dirty red water into a plastic cannister while the team took measurements with the instruments. When he handed the sample to Charlie Howard in the Zodiac the captain took a whiff: "Mmmmm," he said, grimacing.

The crew continued to dive and take samples across the Galapagos over the next few days. "Some of the dives were just magical," said Juan Enriquez, who had rejoined the expedition. "Just wonderful stuff." This included a dive the day after the mangrove sample amidst a mess of iguanas (yes, "mess" is the scientific term). The team saw them sunning on some rocks near *Sorcerer II,* and then underwater. "These iguanas, they look very slow on land when we find them," said Craig, who had just emerged from the water after a dive. "They sort of scamper away. But they are the most graceful animals."

The team kept collecting samples. "A shallow reef sample was taken in Devil's Crown," reported Dill in the trip's log, "a partially submerged crater that is home to scores of fishes, sea lions and rays. We enjoyed the distractions of sea lions while trying to collect our sample. People commented that it was like having a litter of mischievous puppies in wet suits trying to grab our attention the entire time."

On February 17, *Sorcerer II* returned to Floreana. A team hiked into the interior to collect a sample from a freshwater pond brim-

Marine iguanas (*Amblyrhynchus cristatus*), Fernandina Island, Galapagos Islands, Ecuador, February 2004.

ming with pink flamingos. As Jamie Shreeve wrote in *Wired:* "To get a sample from a . . . pond on the island of Florian [Floreana], Venter, Hoffman, and others lugged thirteen-gallon carboys over a hill to be loaded onto the boat. It was worth it. The hundred-degree water was so full of life that the filters clogged up after only three gallons of water had passed through."[13]

The next day, *Sorcerer II* sailed to Daphne Major, north of Puerto Ayora, to pay a visit to this large, mostly barren rock's only human inhabitants, the famous evolutionary biologists Peter and Rosemary Grant from Princeton University. Since 1973, this British couple had spent six months a year on Daphne Major capturing, tagging, and taking blood samples from the finches on the island, birds that Charles Darwin also collected and studied. These tiny birds became a critical component of Darwin's theory of natural selection since they differ by species in subtle but important ways from island to island, having adapted to hyperlocal conditions.

At first, Darwin thought they were different birds. Only later, back in England, did he realize they were closely related, which contributed to his emerging realization that species are not immutable, as scientists had long assumed. The Grants have contributed their own addenda to Darwin's theory, having carefully observed adaptations in the finches on their tiny island as the birds have reacted to changes in environmental conditions like rainfall and food supply. The couple showed that changes in beak size, for instance, can happen much faster than traditional evolutionary theory says they should. Their work was described in a citation when they won a prestigious Balzan Prize in 2005: "Peter and Rosemary Grant...have demonstrated how very rapid changes in body and beak size in response to changes in the food supply are driven by natural selection.... The work of the Grants has had a seminal influence in the fields of population biology, evolution, and ecology."[14]

The thirteen species of finch that live on Daphne Major include tree finches, ground finches, warbler finches, and vegetarian finches. Adaptations include differences in size, shape, and color, and of course in their beaks, which are specifically suited to the types of food they eat. Incredibly, over just the four decades the Grants have spent on the island, they observed changes to average beaks and body sizes.[15]

In 1994, science writer Jonathan Weiner published *The Beak of the Finch: A Story of Evolution In Our Time*, a book about the Grants and their work in the Galapagos that won a Pulitzer Prize.[16] In it, Weiner makes the case, based on the Grants' observations, that evolution occurs rapidly at times and that lineages change, not always in linear fashion. While some species have persisted for many millions of years in roughly the same form,

The volcanic island Daphne Major in Galapagos Islands, Ecuador.

others, like the finches on Daphne Major can fluctuate in their evolution without settling into stability for long periods of time.[17]

Late in the morning of the visit to Daphne Major, after dropping anchor, Craig and a small team from *Sorcerer II* headed to the island and hiked to the Grants' camp. They met the couple near a small lean-to hut where tarps had been draped over the front of a small cave tucked into the rocks. This is where they lived on the desolate island, sleeping on canvas cots in a primitive abode where they also kept canned goods, ample supplies of water, and equipment for their research. Nearing seventy years old in 2004—both were born in 1936—the Grants wore weathered hats over their white hair. Their skin was brown and wrinkled from years of tanning. "We reached their laboratory-camp," said Craig to the film crew, "and it was very simple, the most basic sort of science."

"The Grants were living on canned tuna fish," remembered Craig, describing how the team had taken them an unexpected treat—a bucket of ice holding a very nice bottle of cold champagne, which they used to toast the couple's work. Photographs of the visit show the couple sipping bubbly and smiling. "They were the warmest people," said Craig, "which is not what you expect going to this little rock."

Peter and Rosemary Grant in their tent on Daphne Major, Galapagos Islands, Ecuador, February 2004.

After champagne, the Grants showed the *Sorcerer II* team an example of how evolution worked on the island by pointing out a very hard fruit that grows there. "The island is in a drought," said Craig, "so you can see that the finches with very large beaks can actually crack through this very hard fruit and get to the seeds and eat them to survive." He added that the idea of "natural selection on steroids" is well known with bacteria. "We have evolution taking place hourly in our lungs and in our guts with bacteria," he said, "and constantly in every sample of seawater we bring in. This is on an entirely different scale than birds and other larger life that Darwin—and the Grants—have studied."

"The most amazing thing was that Peter was getting ready to fly to Boston for hip replacement surgery," said Craig, "and we couldn't keep up with him going up and down the mountains. He was like a billy goat."

⋎

ON MARCH 1, after several days in Puerto Ayora, the *Sorcerer II* team took another land sample in the interior of Wolf Island.

They scaled a steep volcano on another blistering hot day to take a soil sample. "This time of year is marked by large swells on this island that make landing difficult or impossible a large majority of the time," wrote Juan Enriquez in the ship's log. "We arrive at sunrise to Wolf and notice the sea swell crashing into the cliffs that rise majestically from the sea. We circle the island in search of a landing site watching the seven-foot swell crash on the rocky shore. It was not going to be easy."

"The island look wild and uninviting as the waves send up plumes of spray," Enriquez continued. "I saw fear in Jeff's eyes as he listened to the discussions of how we would have to time the landing. There is a 120-foot cliff to scale and the only way up is along a narrow ledge then up a cleft in the volcanic rock. By taking the dinghy and riding the surf in near the cliff, we could time it perfectly and jump off the bow of the dinghy onto the ledge at the edge of the water."

"Everyone is excited and anxious," wrote Enriquez. "Jeff begins speaking of death and crazy people. We support him by discussing the sound he would make when he falls, bouncing off the rocks and into the water. Until we realize he is truly uncomfortable and decide to save it for later. It is the moment for Jeff to jump. With our coaching and cheering Jeff is launched off the front of the boat behind Greg, Simone, and Cyrus. I snapped photos and cheered as they clung to the loose rock and began their climb. With a few close calls, falling rock, and teamwork, they all make it to the top."

"They searched for soil and found enough for a sample before facing the trip back down. To return to the dinghy, they decided to take a more direct route. After the heat of the island, plunging into the cool Galapagos Sea seemed a better idea than scaling down the hot dry cliffs again and one by one they leapt the last twenty feet into the sea."

OVER THE NEXT FEW DAYS, the crew returned to Puerto Ayora to rest and prepare for the next leg of the circumnavigation. They planned to depart on March 16 to cross the South Pacific to French Polynesia some thirty-five hundred nautical miles away. On the last day of their break in Puerto Ayora, Craig sat down with the film crew and reflected on what they had seen one last time before they left.

Sorcerer II anchored off of Isabella Island, Galapagos Islands, Ecuador, February 2004.

A gathering in the *Sorcerer II* main cabin, in Port Ayora on Santa Cruz Island, Galapagos Islands, Ecuador, February 2004. Left to right around the table: Compass Light's underwater director of photography Nick Caloyianis; underwater assistant cameraman Vance Wiese; Craig Venter; Simone, our park-issued guide; Kathy Urpani, cook; Karla Heidelberg; Charlie Howard, captain; Bob Friedman; David Conover; Brooke Dill; and Cyrus Foote, first mate. Standing on the right is the local dive guide.

"We've been out here daily, living with real biology and history," said Craig. "We're also seeing the same exact things largely unchanged since Darwin was here. This helps put the biology in context. Biology in some cases has become extremely sophisticated since Darwin's time; in other ways we are doing the exact same thing that Darwin did. We are going out and observing what is there. We are going out and making those observations with new tools. We can look finer; we can see the genetic code. But [like him] we are largely overwhelmed by the information, and even more so. It's not clear to me that humans have the actual intellectual capability, all the brainpower to assimilate all this information, certainly without the aids of massive computing like we are using."

"We're trying to understand factors, the evolution of genes that led to us, or that sea lion, or my dog being alive and active and responsive to our environment. It's so far beyond our current state of biology, of medicine, of intellectual pursuit that all we can do at this stage is make observations and record them and try and look at the patterns of the new facts that we come up with, and we interrelate to—and maybe we'll come up with some additional principles of biology."

David Conover, the documentary producer and director, asked Craig to sum up the voyage so far.

"These have been some of the most wonderful months of my life," said Craig. "And it's not over yet! Tomorrow, we head to French Polynesia, and we'll continue to take samples every two hundred miles. I'm excited to see what we find."

5

French Polynesia to Fort Lauderdale

Only Craig would think of doing something so grandiose.
—CYNDI PFANNKOCH, FORMER JCVI LAB TECHNICIAN

WHILE STILL IN PUERTO AYORA, the crew in early March prepared for a thirteen-day sail to French Polynesia. Craig, Jeff Hoffman, and Bob Friedman also worked to prepare the sixteen samples taken since they left Panama for shipment back to Rockville to be analyzed.

Originally, the team had intended to send the samples back on a commercial flight. This plan was thwarted by yet another official snag when the team brought the samples to the airport. "After we thought we finally had all the issues resolved with our MOU [Memorandum of Understanding] and sampling permit in the Galapagos," recalled Friedman, "another issue surfaced at the airport." They were transporting samples in a "dry shipper," which is an insulated cryogenic container that contains liquid nitrogen absorbed into a porous lining. The FAA allows such containers on commercial aircraft as long as no free liquid nitrogen is present, and liquid nitrogen must not leak from the container. However, when the airport officials opened the lid of the dry shipper to check the contents, the extreme cold hit the warm ambient air and they saw clouds of water vapor condensing. "The airport officials at Puerto Ayora did not know about dry shippers," recalled Craig, "and were frightened by the vapors coming out of the container."

The issue was resolved when Craig decided to carry the samples—still in the dry shipper, frozen at minus-196 degrees Celsius—back to Rockville himself on a chartered Cessna CJ5. He was flying back to attend the press conference announcing the Sargasso Sea paper about to be published in *Science*. After the brouhaha with the Swedes and some of the foreign scientists working at the Darwin station, and their attempts to thwart the sampling effort, Craig was relieved when the jet soared upward from the islands with the containers of eastern Pacific microorganisms. "I think I had a martini on board the plane to celebrate," he later recalled.

Several hours after leaving the Galapagos, the samples arrived in Maryland, nearly three thousand miles northeast of the islands, where the sequencing and bioinformatic teams were poised to begin their work. On a cold early spring day Craig transported them himself to the temporary lab space his institute was still using in Rockville.

Receiving the samples was Cyndi Pfannkoch, the lab tech and scientist who had headed up the sequencing of the Sargasso Sea samples. Pfannkoch had arrived at the institute in 2003 after working at Celera. Before that, she described herself as "a postdoc without a PhD" working for Hamilton Smith when he had a lab at Johns Hopkins. "My lab training," said Pfannkoch, who had a BS in biochemistry, "was pretty much: 'Here, do this for me.' So you learn how to make things happen."

"Ham thought she was one of the best, if not *the* best, lab assistants he's ever had," said Craig about Pfannkoch. "It was the two of them that made all the libraries for [the human genome project at] Celera. Making libraries is an art, and manipulating DNA is an art. Ham had the best hands in the business for being gentle with DNA. DNA is very brittle when you get large chromosomes. For example, with the first synthetic cell"—which Craig and Ham Smith created in 2010—"we couldn't pipette the DNA, because just the act of bringing things in and out of the pipette damaged it. That's how you shear DNA to get smaller pieces, so being skilled at this is no trivial matter."

Wired reporter Jamie Shreeve met Cyndi Pfannkoch in Maryland soon after he returned from his stint on *Sorcerer II* in 2004. He described the process he observed in the lab in Rockville. "The filter papers are first cut into tiny pieces and placed into a buffer that cracks open the cell walls of the organisms," he wrote, "spilling their contents. Chemicals are added to chew up proteins and leave just the DNA, which is spun out of the solution. Pfannkoch pulled a vial from a rack and held it up to the light.

'See that white glob down there?' she said. 'That's DNA from the flamingo pond,'" the inlet that Craig, Hoffman and the team had visited in the interior of Floreana in the Galapagos. "'Compared to a typical sample, this is *huge*. I can't wait to see what's in it.'"[1]

Like the protocol that Pfannkoch, Hoffman, and others had developed for the Sargasso Sea samples, the overall goal here was not modest. It was to take a stab at sequencing bacteria on a planetary scale to better understand the underlying interconnectedness of life on Earth—or at least to learn as much as possible from whatever life was contained in a world of water just below the surface.

"There was me and a group of five or six others that would do the extractions," said Pfannkoch. "Generally, we'd start a couple filters in the morning, get them to a certain point, then start a couple filters in the afternoon. The samples would usually come in on dry ice. We'd open the container and get the manifest out, which was the paperwork and legalese to clear customs. It would say: 'Here are our samples, they are not human.'"

"We would take out the filters that were frozen and wrapped in aluminum foil," she continued. "We would make a record of what sample site we got, put it into a spreadsheet, and then make a note of any information that came with the sample— longitude and latitude, salinity, depth, and all the rest—which we'd enter into a spreadsheet, so we knew what we had received. Then we would go to one of our freezers, that were all numbered because, as you can imagine, with so many samples we needed to know which box they were in, and which freezer."

Later, one of the scientists at JCVI or a collaborator would ask Pfannkoch to pull a specific sample or samples from a certain locale, like the Galapagos, and specify what size filters they wanted to have prepped and sequenced (with the filter sizes ranging from bigger to smaller). "We would go and get the right samples," she said, "hoping that the freezer hadn't gone down and that the sample hadn't been moved to some other freezer, so we wouldn't have to freezer dive, which happened on a couple of occasions. Then we'd get the filters out and we'd unwrap the foil packets, take out the right filter, and move it over to our biosafety cabinet and let it thaw."

Once the filter was thawed and cut into tiny pieces, the scientists started the extraction process. They incubated everything at 55 degrees Celsius and then spun the tubes in a centrifuge to separate the DNA from the liquid in the sample.

The next step was to remove impurities with solvents. "Some of the samples could be really goopy and gloppy," said Pfannkoch, "with algae-like goo, like from the Mangrove Pond. Those were really nasty. We also tried to get a description of what the filter looked like. Was it light green? Was it clear? Was it brownish green? There were a couple that had a rather pungent odor, which is why we opened the samples in the biosafety cabinets. One of them I remember smelled of sulfur. Like a match but without the flame."

Next came the shotgun sequencing process and then the reassembly of the DNA fragments into genes, chromosomes, and genomes as described in Chapter 3. The results were sent to public databases like the National Center for Biotechnology Information (NCBI) and the Community Cyberinfrastructure for Advanced Marine Microbial Ecology Research and Analysis (CAMERA). The latter is a digital depository and website developed as a joint project between JCVI and the California Institute for Telecommunications and Information Technology (Calit2), a center at UC San Diego funded by the Gordon and

Betty Moore Foundation.[2] CAMERA included detailed records not only about the DNA of sample organisms but also about where the samples were collected and their salinity, oxygenation, and much more.

Later, when Doug Rusch and other JCVI researchers analyzed the first forty-one samples of the global expedition, all this data provided the basis for their 2007 *PLoS Biology* paper describing what was discovered.[3] Other publications would detail what the scientists found when they cracked open, sequenced, and analyzed this batch of samples—the last of which, sample 41, was taken March 2, 2004 on Isabella Island as the expedition was winding down its visit to the Galapagos.

Sorcerer II anchored in February 2004 off Pinnacle Rock, Bartolomé Island, Ecuador—a volcanic spire prominently featured in the film *Master and Commander.*

Twenty-seven hundred miles away from all those nebulizers, centrifuges, supercomputers, and rows of sequencers, the crew of *Sorcerer II* on March 17 raised the ship's mainsail at around 6:00 AM in Puerto Ayora's small harbor and aimed the bow west.

Pulling away from the last glimpses of land, the crew marveled as a spectacular Pacific Ocean sunrise unfurled, with high, puffy clouds splitting the rising sol into rays of yellows, reds, and purples. Brimming with supplies, the ship was setting out on what would be the longest single leg across open sea during the entire circumnavigation—a roughly three-thousand-nautical-mile jaunt to Fatu Hiva in the Marquesas Islands, the easternmost chain in the twelve-hundred-mile sprawl of isles that make up French Polynesia. Along with Craig, who had just returned from Rockville, and the core crew, Juan Enriquez had stayed for the crossing. Wendy Ducker also joined the crew, rounding out the ship's complement of ten people.

That night, as the ship's log records, the usual dolphins appeared at sunset, and the crew could clearly see Mars glowing deep-red in a sky packed with stars. Later, those on the late shift saw a meteor shower as the ship covered a distance of 214 miles in just twenty-four hours, a very fast exit from the Galapagos. The second day out, Hoffman hooked a three-hundred-pound marlin around 6:00 PM, but there is no record of him landing it, meaning that he didn't. ("I didn't even come close to getting it on the boat," he remembered later. "It snapped the line.")

"We sailed across the Pacific using a route from Galapagos to Fatu Hiva that gets very little transit," recalled Enriquez years later. "Most of the boats take a different route. Most of the airplanes take a different route."

"We were really alone out there," he said. "There was nothing on the radar," and no lights from cities or towns. "It's like you're in one of those sensory deprivation chambers where you go in and you're floating, or like when you meditate. When you sail across long expanses of ocean, if you don't have major storms, it's almost a Zen-like feeling where you just clear your mind."

"But for some stupid reason, I never had any fear," said Enriquez. "You figure everybody is going to do their job. You don't have an option. You're on watch four hours, you're off eight hours. You can read, you can find people who are simpatico, you can watch movies, you can help the cook prepare stuff." During this transit, Enriquez wrote part of his book, *As the Future Catches You*.[4] Published in 2005, the book predicted where innovation, especially in biotechnology, was likely to take us in the near future. Craig wrote several chapters of his autobiography, too, between the Galapagos and Fatu Hiva. *A Life Decoded: My Genome, My Life*, his story about sequencing the first human genome, was published in 2007.[5]

For Charlie Howard, sailing across the open water on stretches like this could be an isolating experience, "but it wasn't too bad," he said. "We had email, we could check it two or three times a day, keep in touch with people. We had satellite communications. Navigation in open water is actually pretty easy. Watching the weather is something you have to focus on multiple times a day. You're always thinking about preparing for changing weather, changing sails, making sure your rigging's right."

"We would try to download satellite images and predict what weather was coming," said Craig, "because large ocean squalls can be terrifying things. When you're out at sea, you're constantly processing new data and information. Is the engine making a different sound? Is there a different color of water? Is there a different current? Is the wind changing?"

On the way to the Marquesas, Hoffman—now designated "science boy"—led the sampling every two hundred miles, the same distance between samples maintained by HMS *Challenger* scientists when they were dredging the ocean floor in the 1870s. *Sorcerer II's* first intake of microbe-laden seawater was brought on board on March 17 with a sample taken two hundred miles out from the Galapagos.

Meanwhile, Craig, Charlie Howard, and the rest were busily navigating, sailing, and generally trying to get everyone safely across this vast stretch of blue water. On average, *Sorcerer II* did between eight and twelve knots, depending on the conditions. "We can motor in flat water at eight knots," said Howard. "When you have some winds that are useful, you motor at 1500 RPM, then you're doing ten knots. If you have good wind on the beam, you can get up to twelve knots."

Sometimes in big swells, Howard would goose the speed by doing a little *Sorcerer* surfing. "We would get these waves that were probably twenty to twenty-five feet," he said. "They had a lot of energy coming out of the wind and, if they were the right direction, you could get on the wave and go with it. But it gets hairy when you've got a hundred tons surfing down a wave. I mean the risk is high if the steering breaks, and you're going to break something or somebody gets hurt. If the shroud breaks, if a halyard breaks, then the damage level goes way, way up. But *Sorcerer's* a great boat, she could handle it."

They motored and sailed the nearly three thousand nautical miles in just over thirteen days, through patches of doldrums and downwind through light trade winds, mostly coming from the southeast. "For the most part, the ocean lived up to its name, Pacific," reported the log. "There were a couple of washing machine days where everyone on board felt a little dizzy and beat up, but by the end of the journey sea legs had grown to the point where most could use the portable StairMaster on deck without any problem."

"At times, in calm waters, we would jump into the ocean after sampling," continued the log. "It is a somewhat eerie and beautiful sensation to swim in a bottomless blue ocean, literally thousands of miles from land. In a sense it is almost like flying. The sea is completely transparent, falling away to the abyss below and one easily loses a sense of perspective or depth. But it is a far more complex ecosystem than it appears at first glance. Gradually one begins to see large translucent flat jellyfish float by. This is just a taste of all that lives here."

Five days out, the smooth, sometimes Zen-like routine of sailing, sampling, writing books, and fishing was interrupted by alarms ringing and smoke pouring out of the engine room. The engine had overheated, even flames were starting to appear. "The belt started to screech and smoke," recalled Howard years later. "I ran into the engine room in the smoke to evaluate and shut down the main engine and hit it with a fire extinguisher for the flames. I then came up on deck, knowing how nervous Jeff gets, announcing: 'we are all going to die'—although I was pretty sure we were going to be okay."

Disaster averted, *Sorcerer II* was underway again an hour later, only to have the same thing happen the next day. Something was seriously wrong with one of the major mechanical systems on the ship. "We found out that the idler pulley on the main engine had frozen its bearings," recalled Howard, "which wasn't good."

The discovery that the ball bearings on the idler pulley were shot meant that Howard, first mate Cyrus, and others were in for hours of tinkering. "So, our creative team got to work," reported the log, and "quite a bit of banging and cursing later, the team was able to cannibalize the ball bearings from the tensioner arm, install them into the idler pulley, and rig up a new way of tensioning the belt and off we went."

On March 29, at 6:08 AM, the crew spotted land: the island of Fatu Hiva, a thirty-three-square-mile top of an extinct volcano that is a mix of lush rain forest and grassy plateaus. Its mountains rise to almost four thousand feet, and only about six hundred people live there.

"Our first anchorage is among the most beautiful in the South Pacific," Enriquez wrote in the log. "Huge rock spires dominate the 'Bay of Virgins.' The Lonely Planet Guide describes this island, Fatu Hiva, 'as an island of superlatives: the most remote, the furthest south, the wettest, the lushest, and the most authentic.'"

The team reached Fatu Hiva at 10:00 AM and wasted no time getting off the ship and heading out on land. After helping some local people drag their canoes to shore, the crew decided to bushwhack through the jungle and up a rolling stream to a two-hundred-foot waterfall. They were rewarded with a clear mountain pool to swim in at the base of the thundering water.

The next morning at nine o'clock, *Sorcerer II* departed for the island of Hiva Oa, about sixty miles to the north. This is a larger, more inhabited isle where the ship would take on supplies and passports and permits would be processed—or not, as it turned

Sorcerer II at anchor in Fatu Hiva, French Polynesia, March 2004.

out, as politics once again rudely intervened. Just as they had in the Galapagos, Friedman's team had been working for months with authorities in French Polynesia and in Paris, since these islands were still officially territories of France, to get research permits to take samples. "We had coordinated with the oceanographic station on the island of Moorea, which was further east, near Tahiti," wrote Enriquez in the log, "and they were very excited about researching and coauthoring various papers with the team."

This was, however, not to be—at least not for the moment, as officials in Paris moved to intervene in the granting of the permits, despite the favorable disposition of the local government. The authorities ordered *Sorcerer II* to stay in the port in Hiva Oa to prevent any taking of samples while the situation was being hashed out.

Craig and Bob Friedman contacted friends in the US State Department and the French ambassador in Washington, DC, as well as prominent scientists in France, asking for their help. "The following days had frantic cables going back and forth attempting to clear the issue," reported the ship's log. The State Department urged the French to reconsider. They again said no, and the team, now on Hiva Oa, put all the sampling equipment away. They tried to imagine sailing through the twelve hundred miles of islands in French Polynesia without being able to collect microbes.

"France claimed this was their genetic patrimony," added Friedman, "and their scientists were going to do this. So, this was not so much about bioprospecting, but the idea that they wanted to leave any scientific discovery in their waters for their scientists."

The saga would wear on for several days across multiple islands as the crew's frustration increased. "Upon arriving in Hiva

Oa, the Head Gendarme asked us to wait 45 minutes," reported the log, "while he finished other business and then came out and said, 'sorry, closed. Come back tomorrow. 8 AM sharp.' Next morning, we were told that the High Commissioner of French Polynesia requested we not move the boat pending reconsideration of our sampling application. This was, we were told, an informal, not a written request, and they could certainly not detain us if we were to leave."

But they also said that the government could not guarantee safety if the boat left the harbor. Craig passed this not-so-veiled threat on to friends in Washington, which prompted an official communique from the US Navy to the French Navy explaining that *Sorcerer II* was a US research vessel and under the protection of the US government. As tensions increased, the local gendarme on Hiva Oa felt like he was in over his head and proposed that the ship sail to a nearby island, Nuka Hiva, home to the capital of the Marquesas group of islands—plus an airstrip Craig could use to fly to the French Polynesian capital of Papeete on Tahiti. About three thousand people lived there in small villages and towns set amidst more lush valleys and steep, volcanic mountains.

Sorcerer II anchored inside a collapsed volcano that forms the bay where Nuka Hiva's major town, Taiohae, is located. The island is famous for visits from various nineteenth-century European authors and artists during the era when Westerners ruled much of the world and white men delved into the mysteries and exoticness of places like this with intense curiosity, arrogance, and swagger—and, sometimes, with a dark fascination with the dichotomy between the industrialized West and these supposedly more wild and pristine places and peoples.

Herman Melville's novel *Typee* is set in a valley in the eastern part of the island.[6] In 1888, Robert Louis Stevenson, sailing on the luxury schooner *Casco*, visited here, as described in the first chapter of his autobiographical *In the South Seas*.[7] Coincidentally, the *Casco*, at ninety-five feet, was almost the same size as *Sorcerer II*, although Stevenson's boat had two masts and sets of sails, not one. On May 8, 1903, the Postimpressionist artist Paul Gauguin died here, having spent years painting lush works depicting the island as a kind of Garden of Eden—and including himself, a Westerner, in some paintings as a devil-like figure.

"We explored Gauguin's last resting place," according to the log, "visited some beautiful valleys, [and] re-provisioned the boat with food and fuel. The mountains here are huge and dramatic. Afternoon monsoons soaked several of our unwary crew."

Meanwhile, the permit issue was escalating. The standoff continued as Craig headed to Tahiti to meet with authorities and *Sorcerer II* remained "unofficially" impounded in Nuka Hiva. In Papeete, Craig met with a French Polynesian official named Priscille Téa Frogier, who later became the Minister for Education, about taking samples. After some back and forth between governments, Craig agreed to sign another Memorandum of Understanding, emphasizing again that all data generated would be put in a public database. Even this didn't result in an actual agreement as Craig waited in Papeete day after day for the promised permit.[8]

While officialdom creaked along, the crew explored Nuka Hiva, finally getting permission on April 4 to sail to the island of Rangiroa, about six hundred miles to the southwest. "It took about three days to cover this little jaunt," reported the log. "We stopped at the first atoll, 15 miles of palm trees and coral reef,

Tamaroa. As we dropped the anchor, we noticed a little sticking in the steering which turned out to be 5 of the 7 stands in the steering cable broken. After years at sea the half inch thick steel steering cable was holding together by a couple of small strands." Skipper and engineer Charlie Howard made the repair.

On April 12, Craig got tired of waiting in Papeete for the permits and hopped on a plane to meet the crew on Rangiroa. He was joined by Heather Kowalski, who had winged in from the States, and by Shreeve, who would spend the next few days with the team. Departing from Tahiti, Craig was told by the authorities the permits would be ready soon. "All that was missing was a Paris signature, surely by tomorrow," reported the log. But Craig had been hearing this for days now and no permit had materialized, even after a team of local scientists in Papeete joined in his Memorandum of Understanding request. They offered to provide a mooring for *Sorcerer II* and accommodations for the group at a nearby biological station on the island of Moorea.

With Craig back on the boat, the crew, unable to take samples, diverted themselves by diving at a spectacular site called the Blue Lagoon. Here they saw a "massive Napoleon fish, half the size of a compact car, large tuna, many turtles, schools of Wahoo and smiling barracuda." Afterward, Craig enticed Shreeve into a game of hearts and convinced the newbie that he was drunk and not a very good player before going in for the kill. With no science to do, the dives and vicious games of hearts continued, with a toast in the Blue Lagoon on Easter Sunday that featured "Craig's fine champagne" and a meal of "fresh coconut ceviche."

"We left looking longingly at the Blue Lagoon," continued the log the next day. "What a wonderful site this would have been for a water sample we thought as a beautiful rainbow lit the lagoon. Alas, it is unlikely Polynesians will have access to this data or knowledge soon. Imagine someone coming to your town and offering to build you a free library and stock it with data and you saying come back later, time and again. France and Polynesia will not have access or results from their own waters."

The ship headed from the Blue Lagoon to Papeete, where "tomorrow" for the permits still had not arrived. Frustrated yet again, Craig decided to take up the offer of a stopover on nearby Moorea, and *Sorcerer II* headed there April 15. That afternoon they eased into the small bay where the Richard B. Gump South Pacific Research Station is located. Tucked into a thicket of palm trees, it features the distinct sloping roofs that are common in Polynesian architecture. The Gump Station is a collaboration between the French Polynesian government and UC Berkeley, and was at that time conducting a wide range of studies, including projects focusing on the health of nearby coral reefs and a study on the genetics of local islanders.[9] The station hosted a "science day" on the *Sorcerer II* project including a talk by Craig that was attended mostly by scientists and students.

In mid-May, after a long break in Moorea, "tomorrow" came at last and the permits were finally granted. On May 16, the ship departed Moorea and took the first sample in French Polynesia, sample 48, in the inner lagoon of Cook's Bay, and another, sample 49, outside the bay's reef. The team then set off to two nearby atolls, Takaroa and Rangiroa, to explore their lagoons.

On June 8, after another provisioning stop in Papeete and "topping off the fuel," *Sorcerer II* departed for Bora Bora, its last stop in French Polynesia before again heading out to open sea.

Craig Venter diving with sharks at Rangiroa, French Polynesia, April 2004.

The bubbles are one indication of anxiety as team members wait their turn to exit the water on a shark dive, at Rangiroa, French Polynesia, April 2004.

Sadly, the permissions had come so late they were only able to take in four samples from the territorial waters of this sprawling chain of isles.

⋎

ON JUNE 9, *Sorcerer II* left French Polynesia and spent three days sailing to Rarotonga in the Cook Islands, an independent country with ties to New Zealand. The island was "not far geographically, but it's very different, being a colony of New Zealand instead of France," reported the ship's log. "Suddenly everything was English (this took some getting used to since we were too familiar to speak our minds with no one understanding). Our stop had to extend slightly due to a busted seal in the main engine pump. After a two day wait, we received the new part from New Zealand and were up and running again!"

By now, seven months after departing Annapolis, the daily rhythm of the *Sorcerer II* expedition was all about watches, samplings, downtime, sleep, scary movies, and endless games of hearts as the ship passed through the Cook Islands and onward to Tonga, where it arrived on June 22. The authorities there let the crew take samples but insisted that a government representative always stay with them. The official assigned, named Apai, "was a soft spoken, kind man," said the log, "who specialized in water quality for the whole Island group."

On July 4 they reached Fiji, a nation of 330 islands that gained independence from British rule in 1970. The navigation here was tricky, with reefs hidden everywhere just below the surface. "When we were still ten miles out from land, we noticed two fishing trawlers that seemed to be anchored," reported the log. "As we got closer, we realized that they were recent wrecks that

Erling Norrby and Craig Venter sailing from Fiji to Vanuatu, August 2004.

were high and dry." After checking into customs and immigration in the small town of Levuka, the team started taking samples as they sailed from island to island in the Fiji archipelago. Craig, who had taken a quick trip back to Rockville, returned with virologist Erling Norrby, the former Secretary General of the Royal Swedish Academy of Sciences.

On August 4, the crew arrived in Vanuatu, a country once jointly administered by France and Great Britain that gained its independence in 1980. Its islands are part of the archipelago of the New Hebrides in the Pacific region of Melanesia. "Vanuatu is a very rural country made up of a string of 10 medium sized islands and 70 small islands strung over 300 miles," reported the *Sorcerer II* log. "It seems totally unaffected by western influences.

The capital city is Port Vila and has a population of 36,000. The economy is mostly agricultural with a first-rate market which runs in the center of town 6 days a week. The cultures and languages of the surrounding islands vary hugely, with a mixture of around 80–90 different languages."

The first island they reached in Vanuatu, on August 4, was Espiritu Santo. In 2004, this was a sleepy island. During the Second World War, the island was used by the US Navy as a major supply base in the war against the Japanese. James Michener drew on his time as a lieutenant commander there to write *Tales of the South Pacific*, for the book that inspired the Rodgers and Hammerstein musical *South Pacific*. Scuba divers still love the island because, after the war, the American military dumped most of its remaining equipment and garbage into the sea at what is now called "Million Dollar Point."

Craig explained how the point earned its name: "The Americans were pulling out of Vanuatu at the end of the war. It was returning to French governance. The US wanted the French to pay some money for all the equipment and trucks and jeeps and stuff they were leaving behind. And the French said, 'you're leaving them anyway. Why should we pay you for them?' So, the Americans had this fight and built this long cement pier. It's called Million Dollar Point because they drove all the tanks and trucks and jeeps into the ocean."

The largest dive-able shipwreck in the South Pacific is also there, the SS *President Coolidge*, a 654-foot luxury passenger liner that was converted by the US Maritime Commission during the Second World War to carry troops and freight. "On October 26, 1942, as she entered Luganville harbor, she struck an American mine amidships on her port side," reported the *Sorcerer II* log, recounting the history. "A second explosion was heard soon after and it is believed a boiler exploded. Captain Henry Nelson ran her ashore on the reef in an attempt to save the 5,150 troops, 290 crew, and cargo. Amazingly, only two men lost their lives, but this beautiful vessel slowly sank to where she is now resting on her port side with her bow in 65 feet and her stern in 240 feet. Her cargo was not removed and much of it was still there for us to explore."

"After a day of exploring these historical sites and the very sleepy town of Luganville," the log continued, "we were ready to make an overnight passage to Port Vila on the main island of Efate. We paused to pick up our water sample about a hundred yards off the Coolidge wreck. As we left, curious children came for a visit on their homemade raft made from a scrap of Styrofoam."

According to Craig, the area is also known for its sorcerers and magic, which provided a special bond to locals when they found out the crew was sailing in a boat named *Sorcerer.* "The ceremonies and customs have changed very little over time and rituals are strictly adhered to," according to the log. "We were careful to move within the villages with respect for their way of life and the women of *Sorcerer* were careful to wear clothing that covered their legs and shoulders." Brooke Dill also wrote in the ship's log about her personal experience of visiting this island:

> As an American woman, I found the Vanuatu way of life different than anything I had ever experienced. When stepping on shore for the first time, the first inhabitant to greet me was a feral pig tied to a tree. We surprised each other and I walked well around this piglet so not to stress him too much. I soon found out that this pig really did have right of way. Pigs are above women in the commu-

nity hierarchy (men then pigs then women). Yes . . . this piglet had something on me. I also found that I could be bought for a few pigs.

The visit to Luganville had special significance for Craig: his father had been stationed here during World War II. John Venter was a staff sergeant in the Marines, Craig said, although he didn't know much more about what his father did during the war. "He never talked about it," he said. "My mother joined the Marines at the same time he did to see the world, and she got stationed in San Diego. They met in Oceanside at Camp Pendleton."

"Exploring the shores and waters is a constant reminder of the thousands of American soldiers stationed there," reported the ship's log. "While beachcombing, WWII-era Coca-Cola bottles litter the undergrowth of the jungle and medicine bottles are found in the sand as a reminder of the extreme illnesses, such as malaria, that they endured. Military equipment rests silently below the surface of the ocean where it was abandoned, creating a home for octopus and fishes of all shapes and sizes. It was a brief glimpse of the life [Craig's] father led while stationed in the South Pacific."

When *Sorcerer II* left Vanuatu on August 20, its crew was down to only five people. Craig had flown back to the States with his son, Christopher, and Erling Norrby. Juan Enriquez had returned home to Boston with his ten-year-old son, Nico, who had been along for this portion of the trip. The ship's complement was now down to Charlie Howard, Jeff Hoffman, Cyrus Foote, Brooke Dill, and the cook, Tess Sapia, for the sail to New Caledonia and then onward to Brisbane. The plan was to wait out the cyclone season in Australia for the next few months and to put *Sorcerer II* in dry dock for repairs.

While the team was in Vanuatu, Shreeve's article about Craig was published in *Wired*. Titled "Craig Venter's Epic Voyage to Redefine the Origin of the Species," the article was "mostly about Craig and the expedition," noted the log, "with little snippets about the crew and the boat etc.... All in all it is a good read and supplements our logs quite nicely." The article starts like this:

> **Picture this:** You are standing at the edge of a lagoon on a South Pacific island. The nearest village is 20 miles away, reachable only by boat. The water is as clear as air. Overhead, white fairy terns hover and peep among the coconut trees. Perhaps 100 yards away, you see a man strolling in the shallows. He is bald, bearded, and buck naked. He stoops every once in a while, to pick up a shell or examine something in the sand ...

This is, of course, Craig doing his thing, although Shreeve noted that the man in question wasn't there just to have fun.

> What separates him from your average 58-year-old nude beachcomber is that he's in the midst of a scientific enterprise as ambitious as anything he's ever done. Leaving colleagues and rivals to comb through the finished human code in search of individual genes, he has decided to sequence the genome of Mother Earth.[10]

⩔

THE PARED-DOWN CREW rose early on August 20 to head west for the short, 262-nautical-mile jaunt westward to the French territory of New Caledonia. The crossing and sampling went smoothly, reported the log, except that the warm, tropical weather they had grown used to abruptly shifted to colder temperatures: "A brisk wind had us running for warm clothes and looking for the heat of the sunshine we had become so accustomed to." As Hoffman later recalled, "It was like winter all the sudden came in."

On August 21, *Sorcerer II* approached the barrier reef surrounding the main New Caledonian island of Grand Terre. At 10:00 AM it passed through the reef's entrance, spending the rest of the day navigating through a complicated network of reefs and islands. When the crew finally arrived in the capital city of Noumea, they were greeted by the dockmaster with a hearty "bonjour!" As the log reported, "We found ourselves a bit off balance walking ashore to a busy French city with French cuisine and markets and carnival rides brightly lit right next to the marina. This was drastically different from the quiet, secluded areas we had been sampling for the previous months."

After a few days of cleaning up and reprovisioning *Sorcerer II,* they headed southeast to gather a sample, hoping the temperature would rise.

It did not. In fact, it felt more like an Alaskan fjord than the South Pacific. They reported seeing, as they sailed to the island of Baie de Prony, "spectacular rolling red mountains and pine trees next to palm trees."

As they tooled around the islands of New Caledonia, Hoffman took a total of three samples. Then, on September 11, they returned to the capital, where Craig rejoined the expedition along with Dave Kiernan and Olivia Judson, an evolutionary biologist from London and author of the bestseller *Dr. Tatiana's Sex Advice to All Creation.* Back in Noumea, the team visited with local scientists from the French National Research Institute for Sustainable Development, who invited Craig to give a presentation on genomics and the *Sorcerer II* project.

On September 17, *Sorcerer II* departed New Caledonia, headed to Australia. "The ship surged along at 11 knots with the engine off," said the ship's log. "The fishing lines were out and for dinner we all enjoyed what we thought was fresh Wahoo but in hindsight we now believe was a 'Spanish Mackerel.' By the next morning the wind had settled down to a light breeze and all the crew except for Jeff were having strange sensations ranging from tingling in the extremities, aching bones and joints, fatigue, upset stomachs, hot–cold touch reversals, and bubbling sensations on the tongue when drinking water. We concluded that we had eaten fish containing the toxin ciguatera and could expect to suffer the symptoms anywhere from a few days to many years." Fortunately, most of the crew felt better by the time they reached Australia.

▿

AT 7:00 AM ON SEPTEMBER 21, the crew came in sight of Brisbane, a city of tall skyscrapers with a population of nearly two million people. By far, this was the largest city they had seen since leaving Panama City nine months earlier. "We were amazed at seeing this modern city as we motored in," remembered Hoffman as the expedition prepared to spend the next eight months down under, sailing and conducting extensive repairs and outfitting for the final legs of the circumnavigation that would take them across the Indian Ocean to Cape Town, South Africa, and then across the Atlantic back to the United States.

First, though, *Sorcerer II* needed to navigate the Brisbane River, where the ship would be docked for the next month. The city was laid out on either side of this meandering waterway, interspersed by bridges. "After four hours of navigating sandbars and shifting channels up the river we arrived at the quarantine dock," reported the log, "where an inspector confiscated all our plants, vegetables, fruits, dairy and poultry, and half our frozen meat. Customs and immigration then greeted us and officially welcomed us to Australia. We felt special when we realized that we were the largest sailing yacht in the whole city and residents were amazed to hear that our mast was too tall to fit under any but the first bridge in the city, leaving us only one place to dock. Fortunately for us this is a very nice, secluded location at the center of the city"—an inner suburb called Kangaroo Point—"and for the first time since leaving Florida 9 months ago we can plug into reliable shore power, and all get connected with cell phones and internet again."

Australia was designed to be a long layover to wait out nasty weather in the southern Indian Ocean and to repair and spruce up *Sorcerer II*. Heather Kowalski—who arrived in Sydney from the States—also worked with Craig to set up a brisk lecture and media tour in a country that welcomed him, seeing his maverick nature as almost Australian.

Torsten Thomas, a microbiologist who helped the team select sites in Australia, saw a personality fit, as well: "I think he was appreciated in Australia because we're fairly easygoing and we like our fun, and I think he enjoyed that." Torsten is now a professor in the School of Biological, Earth, and Environmental Sciences at the University of New South Wales, and director of its

Centre for Marine Bio-Innovation. "I think Craig likes Australia a lot," he continued. "He once said to me that he felt more at home in Australia than many places. I mean, he's a bit of an eccentric and kind of does his own thing. He has a sense of humor and likes to have a good time." The team also organized collaborations with the University of Queensland, University of New South Wales, University of Melbourne, Australian Institute of Marine Science, Commonwealth Scientific and Industrial Research Organization, and Great Barrier Reef Marine Park Authority.

While at the dock in Brisbane, the crew also had to deal with a suspicion on the part of the Maritime Border Command, essentially Australia's Coast Guard, that there was illegal activity on board *Sorcerer II*. "I had just arrived back on the boat from the USA and was sitting on the deck having a beer," recalled Craig, "when I noticed a police boat with a full SWAT team pull into the slip next to us. It became clear that they were coming to see us when their captain, a woman about six feet tall, came up to me and said they needed to board as they had a report that we were running a meth lab onboard."

Craig kept his cool. "I said that we were a US research vessel and they did not have the right to board, but if they agreed to take off their boots and ask a crew member to open any compartments they wanted to see, then they could come aboard. The captain agreed and I accompanied her and some of her team to the ship's interior, where it became immediately obvious that we did not have a meth lab. One of the officers said, 'Somebody's pretty pissed off at you guys.'"

As it turned out, it was rumoured to be someone's jilted ex-lover who was to blame for the kerfuffle. "So apparently the boyfriend sent the police around to get even. The SWAT team agreed to pose for photos with us after they refused my suggestion that they should arrest Jeff."

On October 27, *Sorcerer II* departed Brisbane for Sydney. Upon arrival, the crew motored over to the Australian National Maritime Museum to meet up with Craig, Heather Kowalski, and Craig's niece Melanie Venter, who was helping to manage the PR tour. From there, Craig set out on a ten-day Aussie tour of press briefings, lectures, and meetings. "Basically, Craig was a rock star there," recalled Hoffman.

In Sydney, Torsten Thomas showed the team around and helped to select sampling sites during their stay in Australia. Originally from Germany, Thomas came to Australia to get his PhD in the late nineties and has lived there ever since. "I found Craig's project quite interesting," he said. "At the time, everyone was excited about DNA sequencing and the technologies being developed. The idea of applying this to large samples of microbes was quite remarkable and revolutionary at the time. I had been working with my colleagues for many years trying to understand chromosome sequencing in the marine environments, and having this technology applied to the samples was quite eye-opening—then and now."

Thomas remembers Craig giving a couple of lectures at his university. "He was quite entertaining in his presentations," said Thomas, "and he really wanted to drive home how to use genomics in the ocean and beyond. He was very fired up about this."

"Since then, we've published a number of papers," he added, "with the first ones coming out of the Global Ocean Sampling expedition.[11] We've been able to show that there's a lot of unexpected and undiscovered diversity out there in the marine samples."

Starting in Australia, Craig became something of a mentor to Thomas. At the time, the young researcher was trying to decide whether to work in the private sector or pursue the academic life, and he expressed some worries about the competitiveness in universities. "Academia is always a bit risky, to build a career, to be a professor," said Thomas. But Craig reassured him by saying, "Look, we'll do this big project together and then you're going to be fine."

"Craig is somebody who just says, 'Follow your vision and things will work out.' And he was absolutely right because me getting involved with the *Sorcerer* project was hugely helpful to my career. A lot of this data that we generated really helped me accelerate my subject and career."

While in Australia, Craig also met with the New South Wales government to ask for financial support for the expedition and for the science. "This was difficult," said Thomas. "Craig is straightforward and direct, and the New South Wales government was a bit convoluted in their approach. I think Craig doesn't have much time for little political games. He just wants to get things done—ideally, yesterday. People who can live with that work well with him. They get a lot of things achieved. But it's certainly a high-pressure environment."

"The politicians threw up a lot of red tape," Thomas continued, "but they did end up supporting the project, even if they had a hard time understanding his vision." Eventually, some funding for joint studies between JCVI and the University of New South Wales came from the Australian Research Council, the equivalent of the National Science Foundation in the United States.

"Before Craig came, we had been looking at the micro-diversity of marine organisms," said Thomas, "like what kind of microbes and what kind of diversities have been found about microbes that are living in symbiosis with sponges and seaweeds and other sorts of important organisms. And I think that's something that Craig hadn't really considered as being such a huge barrier to understanding microbes. Because all life in the ocean lives in symbiosis with microorganisms, and you need to understand what those microorganisms do: the diversity, the stability, the function, et cetera, to really understand how those higher organisms function in their environment."

"We pointed out to him that there is a lot of interesting science that can be done in this symbiosis area," said Thomas. "And he said, 'let's see whether we can get some joint funding in for this.'" That funding included not only the ARC but also some money Craig helped secure from the Moore Foundation's Marine Microbiology Initiative, which allowed Thomas and his colleagues to drive a lot of research. "This is an area that I'm still working on almost fifteen years down the track."

⌄

JEFF "SCIENCE BOY" Hoffman and his team snagged thirty-four samples during the ten months in Australia. The first six were taken in the Sydney area, at sites chosen by Thomas and colleagues. These included Homebush, under the harbor bridge; a mid-bay sample; Bare Island; and a time series at Botany Bay. These samples later resulted in a number of studies on the diversity of microbes, microbial communities in different ecosystems, and some of their functions.[12]

Not all the samples were taken using *Sorcerer II*, which stayed in dry dock for weeks of repairs and retrofitting. While this was going on, Hoffman and his team took samples in salt ponds, Melbourne, and other locations in southeastern Australia. The team used a hundred-pound portable filter contraption they dubbed "Beast I" and traveled in a four-wheel-drive, diesel-fueled Toyota Land Cruiser they named "Beast II." The Land Cruiser had been converted into a camper with two full beds, a refrigerator, a sink with running water, and a gas stove. The team added two two-hundred-liter carboys, an air compressor, pumps, a sample cooler, tubing, and, of course, Beast I.

The *Sorcerer* "road tour" left the team's base in Brisbane on February 18, 2005, "with dancing Elvis on the dashboard, 20 gigs of music on the iPod, and magnetic *Sorcerer II* expedition signs on the sides," wrote Hoffman in the road tour's log. "First sampling stop was Lake Tyrrell in a small town called Sea Lake. To get to this sleepy little town, drive 12 hours to Sydney. Then wake up the next day and drive 13 hours due west into the middle of nowhere. Along the way you will almost hit giant kangaroos and have to pull over due to the type of locust storms you read about in the Bible."

"The town of Sea Lake is 10 miles south of Lake Tyrrell and 4 hours north of Melbourne in Victoria," continued the log. "It is the largest lake in Victoria at 20,860 hectares and serves as an end point for the Avoca River and Tyrrell creek system. The shallow lake dries up in the summer months leaving salt on the lakebed from the ground water. This salt is harvested, purified, and sold for industrial use by Cheetham Salt Limited."

"Entering the lake near the Cheetham factory you see a pile of salt four stories high and 200 yards long next to a few small buildings. Looking across the massive lakebed, all there is to see is salt. It looks like a giant field of snow." Cyrus Foote and Jeff Hoffman were greeted by Terry Elliot, the site supervisor. He informed them that it might be tough to find an area to sample, due to the hundred-degree heat that had evaporated most of the lake. "After driving on roads made out of salt, a small pink / purple pond was found for sampling," continued the log. "The pond was about 5 times the salinity of the sea and was teaming with microorganisms of all sizes. The usual amount of 200 liters was not necessary for this sample. 20 liters and 3 filter changes per size fraction was plenty to work with. Each filter size had a unique reddish to pink color."

The next stop on the road sampling trip was Melbourne, meeting up with University of Melbourne collaborators Mike Dyall-Smith and Peter Janssen. "Peter studies phylogenetic diversity of soils," said the log, "and has sent soil / DNA from one of his sample sites to the Venter Institute for environmental shotgun sequencing." The team took five samples over the next six days, including one that the scientists processed at the Royal Melbourne Yacht Club. When they realized they had interrupted a wedding at the club with their loud air compressor, they made sure to turn it off during the vow ceremony—and then joined the party.

Hoffman and Foote's next foray in Beast II was in Tasmania, 230 miles from Melbourne. To get there they took a ferry, the *Spirit of Tasmania,* to Davenport, in northern Tasmania. After arriving, the team drove south to the state capital of Hobart for a meeting at an office of the Commonwealth Scientific and Industrial Research Organization, where local scientists helped them pick out the best sampling sites. "Two sites were chosen in

Tasmania," wrote Hoffman. "They were the southernmost samples taken on the whole trip and not surprisingly had the coldest water sampled since leaving the east coast of America." In the end, Hoffman and Foote traveled four thousand miles in Beast II, plus the thousand-mile round-trip journey on the *Spirit of Tasmania*.

When Hoffman and Foote returned from *way* down under, *Sorcerer II* was back in the water and ready to sail north to the Great Barrier Reef. With the ship repaired and provisioned, they departed Brisbane on April 22. "The first stop was Heron Island where they met with University of Queensland researchers that have facilities on the island and to take a sample from the surrounding reef." They then took samples at Britomart Reef, Heron Island Reef, Dunk Island, High Island, and an Australian Institute of Marine Science Restricted Research Zone. After Heron Island, the expedition stopped in the Whitsunday Islands, 560 miles north of Brisbane, for a media stop and to meet with more collaborating scientists. The team finished in Australia, with more samples taken near Darwin and the Great Barrier Reef, before aiming the great vessel west yet again toward the vastness of the Indian Ocean and Africa, almost seven thousand miles away.

WHEN *Sorcerer II* departed Darwin on July 21, 2005, pounding through choppy, gray-green waves, a brisk wind was whisking the crew toward a routing that would take them just south of one of the old spice routes from Jakarta to the Seychelles—and, later in the voyage, into trouble with gun-toting commandos.

Their first stop was Christmas Island, roughly fifteen hundred nautical miles from Darwin. Promptly upon arrival, everyone on board became ill. "We all got land sick," remembered Hoffman. "I don't know why, but we all got off the boat, we were walking, and we all started swaying like we're drunk. Thankfully, it quickly passed." Next up was Cocos Keeling, about 540 miles from Christmas Island. On one of the few inhabited islands in this cluster of atolls, Craig and his son, Chris, flew in with Juan Enriquez to join the crew. "What I remember most about Cocos Keeling was that it has the clearest water I'd ever seen," said Hoffman. "I remember we were dropping anchor and I was screaming at Charlie to stop because I thought we were going to hit something. And there was still fifty feet of water below us."

From Cocos Keeling, *Sorcerer II* kept going, sailing nearly fifteen hundred miles to Chagos Island in the Chagos Archipelago, some twelve hundred miles southwest of India. Chagos is near Diego Garcia, home to a US Air Force base that supports a squadron of B-52 bombers, in territory that is controlled by the British. It's a spot where the British and the Americans are known to get a bit testy and are prone to hassling boats now and then, if they wander too close. The islands were uninhabited until the French established copra plantations there, using slave labor, in 1793—copra being the coconut flesh used to extract coconut oil. Chagos, Diego Garcia, and other islands in the archipelago became British after the Napoleonic wars in 1814. Over the years, the British and later the Americans built military bases on these tiny spits of land. Both still consider the islands to be strategically vital from a geopolitical perspective. In 1965, the islands were recast as the British Indian Ocean Territory, with a promise by the British government that they would be ceded to nearby Mauritius when they were no longer required for military purposes.

A half-century later, the islands have not been ceded. Instead, in the late sixties, London ordered the removal of the islands' populations. This command was not well received by the locals, who were rounded up by US troops from the base on Diego Garcia in the early seventies and shipped off to Mauritius. The Chagos people have challenged this in British courts and are still fighting the US government for the right to return to their islands. *Sorcerer II* had its own challenges with the British authorities while visiting Chagos.

"At the time, we of course knew about the B-52 base on Diego Garcia," remembered Craig, "and also that the sailing guides highly recommended a stop in Chagos, as it had a great anchorage and one of the most beautiful atolls. Being about halfway across the Indian Ocean, I thought it would provide a nice break to briefly stop there. I notified the State Department that we planned a brief stop at Chagos, and that no samples would be taken. So we arrived and anchored at a place where there were several boats also at anchor. We asked one of them if there was a charge for staying there. We were told that a British rib would come by each evening and collect a small fee and that the stay was limited to one week."

"After we arrived, we were relaxing in the cockpit enjoying, for once, a calm sea," continued Craig, "when we saw a black rib stop briefly at each boat until they got to *Sorcerer*. That's when things changed rapidly. There were six in the rib and, as they approached, we could see they all had their weapons drawn. They had no flag, no uniforms, no insignia to indicate any country, military, or police affiliation. A woman who seemed to be in charge announced that they were going to board us, which they did with their weapons still drawn. She said that they had been looking for us and had hoped that we would not stop there. I indicated that I had notified the US State Department that we were making a brief stop and not taking any samples. She demanded to see our passports, which she took and refused to return. She said we were there illegally and that they could confiscate the boat and arrest us. They left an armed guard on board and indicated that they would return the next day when they had instructions."

"*Sorcerer* was equipped with modern high-speed satellite telephone and data," said Craig, "so I immediately went to my cabin and telephoned a friend, who happened to be the US Ambassador to the United Kingdom. He took my call immediately and I explained how we had been boarded at gunpoint by six presumed British military, and that they took our passports and left an armed guard on board. He said he would look into it immediately and to call him if anything else happened. A short while later he called me back and asked if we had any dive equipment on board. I indicated that we had numerous scuba tanks and dive gear. He said that they had a law from the 1800s requiring a government permit for a sailing vessel to possess dive equipment. Apparently, this was the only law they could find that we had violated. The ambassador said they would return with a document for me to sign, saying I was in violation of their law. And if I signed it, we could leave."

"When the boarding party returned, it was in a British Navy rib with a large, formal British flag on the back. All were in dress military uniforms with weapons in hand. The woman in charge and one other accompanied me to my cabin, where I called the ambassador. The woman officer said that I could sign the document and she would return the passports and we could leave, or she would leave it with me and return in one week to see if I would sign it then. The document also bound me to secrecy about the whole event. Apparently, that was their greatest fear—that due to my fame, it would attract attention to the dispute over the islands."

"I explained to the ambassador that I would sign it, but that I was signing it under duress due to the fact that two armed soldiers with weapons out were standing in my cabin and threatening to hold us hostage if I did not sign. So I signed the document, our passports were handed back to me, and we were told to leave immediately. I thanked the ambassador, the boarders left *Sorcerer*, and we immediately pulled the anchor and left the atoll."

"We were about two miles out to sea from the island when my satellite phone rang," said Craig. "It was the ambassador. He instructed me to stop the boat immediately or we would be fired on. We were under power and quickly stopped the boat. The ambassador indicated the British government would not accept a document signed under formal declared duress. The ambassador said: 'I cannot tell you what to do but I want you to think very, very carefully before you answer the next question because the consequences could be very serious: *Were you under duress when you signed the document?* Now I *was* really under duress, because if I answered wrong, we could be arrested, and the boat impounded. I let a full minute go by and then said, 'Mr. Ambassador I was not under duress when I signed.' He said, 'good, that is the right answer, you are now free to go. Have a safe rest of your voyage.'"

From Chagos, *Sorcerer II* headed for the Seychelles, 978 miles to the west. "The trip across the rest of the Indian Ocean to the Seychelles was quite rough," remembered Enriquez. "As we crossed below the Arabian Peninsula and approached the coastline of Africa, tensions on the boat ratcheted every day as we entered a lawless area of the sea. At night, any blip appearing on the radar had the potential to be a pirate ship. At the time, large cargo ships were being systematically boarded, and crews held for ransom. We read the daily pirate risk dispatches very carefully. Anyone coming too close, especially at night, led to an all-crew watch on deck. Parts of the boat were somewhat protected from bullets, but all in all we were sitting ducks: a large, wealthy vessel in some very tough waters."

"So it was an enormous relief when we arrived in the Seychelles," continued Enriquez. "Suddenly, pristine white-sand beaches and otherworldly boulders allowed picnics, laughter, and, after the long dry spell of open-ocean passages, a wee bit of tequila."

"The biology of the Seychelles is otherworldly," said Enriquez. "Starting with telephone wires covered by spiders the size of both of one's hands—they were harmless, said the locals. After a few blissful days hiking and surfing, most of us left for the US, and *Sorcerer* began heading south toward some of the most dangerous waters on Earth—not from humans, but from a large population of Great Whites and from hundred-foot rogue waves known to shallow entire oil tankers."

During a dive with a local research team studying whale sharks, Craig found himself directly in front of a large whale shark with its mouth wide open. "I quickly kicked my fins hard to keep from being drawn into its mouth," said Craig—while also noting that whale sharks were a great connection for the expedition, since this huge fish feeds on some of the smallest organisms in the ocean, including microscopic plankton and tiny shrimp and fish.

Y

ON OCTOBER 4, after a seven-day sail from Reunion Island, *Sorcerer II* reached the African Continent at Port Elizabeth, South Africa. After ducking into the port to avoid foul weather for a day, the expedition rounded the Cape of Good Hope, a bleak, rocky tip of Africa originally called "Cape of Storms" by its Portuguese discoverer, Bartolemeu Dias. Here, the ocean crashes into rocks churning with foam and low, brown cliffs are topped with a sparse burr cut of green foliage.

Sorcerer II was escorted as usual by dolphins as it sailed into Cape Town, a city of gleaming white towers and Cape Dutch buildings tucked into a shelf of land beneath the brown, rocky, aptly named Table Mountain, a ridge that rises above the city. The expedition motored into the port, docking at the V&A Waterfront Marina, which became a base for Craig to give lectures and for the team to check out the city and take a safari. Craig also rented a Harley-Davidson and rode from the city to the Cape, while the team took a sample of Cape water, before *Sorcerer II* departed on November 16.

Y

FROM November 16 to December 15, 2005, *Sorcerer II* sailed across the South Atlantic Ocean, traveling an astonishing 5,682 nautical miles in twenty-nine days from Cape Town to Antigua in the Caribbean. "We set off along the west coast of Africa using favorable trade winds to make good speed," remembered Erling Norrby, who joined the expedition in Cape Town. "It was beautiful to see the birds gliding along the powerful waves, using the upwind they created." Norrby added that Jeff finally caught a big fish outside of Cape Town, when one of the fishing rods mounted in the stern with a line out let loose an alarm that a fish had been hooked. "After heavy work for about half an hour, Jeff and Charlie managed to land a sixty-pound yellowfin tuna fish," said Norrby. "Its meat was delicious, and we enjoyed large quantities of it prepared in different ways."

Sensing home and the end of the voyage, the weary crew didn't dawdle, although they did collect nineteen samples and made time for a stop at St. Helena. The crew visited the last home of Napoleon, who was held here as a prisoner of the British for the final six years of his life. *Sorcerer II* also briefly stopped over on the tiny isle of Ascension, which was used as a relay station for communications for the Allies during the Second World War.

On the way back across the Atlantic, a troubling equipment problem once again struck, as a hole formed in the mainsail along one of the seams. This meant that Howard had to be very careful using this sail and had to depend more on the motor than usual. A fix would have to wait until the ship's arrival in Antigua.

On December 15, the *Sorcerer II* expedition arrived back in the Americas at Antigua, where it became clear that the ripped sail was beyond the capacity of the locals to repair. "When we got into Antigua and we pulled it down and we took it to the sail-repair people," said Howard, "they looked at it and they rolled their eyes and said they couldn't really do much. So we bundled it back onto the boat—and that's a massive job, to get that sail off the boom and onto the shore, and then get it back on the boat." Craig flew to Antigua to meet *Sorcerer II* and sailed—or, rather, motored—with the ship to his villa in nearby Saint Bart's.

"We sailed over to Saint Bart's the day after Craig got there," said Howard, "with the sail basically dumped onto the deck. And we spent Christmas and New Year's in Saint Bart's with Craig and his friends and family. When we got to Florida, we had the sail repaired. We used it for another four years."

On January 12, 2006, *Sorcerer II* arrived back in the United States at Palm Beach, Florida, 875 days after arriving in Halifax in August of 2003.

⅄

IN THE END, the circumnavigation was even more fruitful than Craig thought it would be, in almost every way: as a "big science" venture, as an expedition that faced substantial hurdles in logistics, weather, politics, and funding, and as a philosophical and even poetic endeavor. By its nature, it asked and tried to answer some of the most basic questions about life and who we are as humans, and also how life is intricately connected on our tiny, blue-green planet. The scientific outcomes added a substantial jolt to the already surprising findings from the Sargasso Sea about the true diversity of life in our oceans. They also proved that epic explorations of life on Earth like the voyages of the *Beagle* and *Challenger* were still not only possible but vital to understanding the secrets of ocean microbiome.

The Global Ocean Sampling expedition helped inspire other "big science" ocean voyages, and contributed to a revival of the grandiose notion that it's possible to do what Craig had said to Jamie Shreeve back in 2004, when the Sargasso Sea paper was released: to think in global terms of sequencing the DNA of the planet. Not literally—at least, not yet—but conceptually, in terms of using science to at least begin to understand how microbes work in huge, global systems to connect and shape life on Earth.

Nor were the expeditions over—not by a long shot—as *Sorcerer II* returned to Newport later in the summer of 2006 to be repaired and revamped once again. Its mission would run for another decade-plus, until the summer of 2018.

Meanwhile, the massive scientific endeavor to understand what the team had collected was operating at full throttle in Rockville. That's where Craig, Ham Smith, Cyndi Pfannkoch, Doug Rusch, and others were processing, sequencing, and analyzing the astonishing treasure trove of samples and data, while teams of scientists pored over what it meant. They would spend the next year-plus—and well beyond that, right up to the present—analyzing the smudgy muck on the filters. An endless stream of papers would start with the special collection that *PLoS Biology* decided to publish in the spring of 2007, even as *Sorcerer II* was again out at sea taking more samples. In the years to follow, the great vessel would plunge through the waters off the Pacific coasts of Central America, the United States, and Canada, and onward to seas as far-flung as the Cortez, Baltic, Black, and Mediterranean. All the while, it would haul in ocean water filled with Earth's smallest organisms, as the adventures and the science continued.

6

Questing Distant Seas (Further Explorations)

*It is a strange thing that . . . most of the mystical outcrying
which is one of the most prized and used and desired
reactions of our species, is really the understanding . . .
that all things are one thing, and that one thing is all
things—plankton, a shimmering phosphorescence on the sea
and the spinning planets and an expanding universe.*

—JOHN STEINBECK, *THE LOG FROM THE SEA OF CORTES*

IN MID-MARCH 2007, *Sorcerer II* was anchored in the crystal-clear waters off the Mexican village of Loreto on the Sea of Cortez. It was a little over a year after the ship had arrived back in the United States at the end of its circumnavigation. Now the team was on yet another leg of the expedition, on the other side of North America, collecting samples along the western coast of this finger of ocean that runs between the west coast of the Mexican mainland and the Baja Peninsula.

During the sojourn on this narrow, luminescent sea, Craig was reading John Steinbeck's *Log from the Sea of Cortes,* which describes the novelist's 1940 voyage with marine biologist Ed Ricketts to survey marine life along the coast of Baja California. "Steinbeck writes about the astonishing life here," said Craig, who often brought along books written by explorers and writers who had traveled to and written about a place *Sorcerer II* was visiting. "It was exciting to follow Steinbeck as we saw the

Jeff Hoffman and Craig Venter taking a water sample in a stream between Loreto and Mission San Francisco Xavier, Baja California Sur, Mexico, March 2007.

Brett Shipe and Charlie Howard hauling a water sample, and Craig Venter and Jeff Hoffman in the background sampling from a salt pond, on Carmen Island, east of Loreto, Baja California Sur, Mexico, March 2007.

places he saw sixty years before our trip, when we were observing a different kind of life than what they were seeing."

"Life was more abundant all around us than anywhere in the world," said Craig about sailing in Cortes. "Blue whales, the largest of whales, would surface and swim right past us, while rays would leap out of the water and seem to fly through the air."

Samples taken included one on March 19 from the small Loreto River that cuts through the village, and another on March 21 from a small salt pond on nearby Carmen Island.

Sorcerer II hadn't stayed still much since the circumnavigation ended in January 2006. "As soon as we got back, we were already looking at how to make a more thorough sampling of the Earth's oceans," said Craig, "because it seemed incomplete to leave out other major seas." The ship had stopped for a six-month overhaul in Palm Beach and later in Newport. In August 2006, Craig had taken *Sorcerer II* up to Cape Cod and to Rockport, Maine. Later, he and his team had returned to Bermuda to collect two more samples, then sailed onward through the Caribbean and the Panama Canal to the Pacific side of Mexico. On this hot March day they had wound up in the sleepy town of Loreto. They planned to keep going until they sailed back south, around the tip of the Baja peninsula, and then up the Pacific coast to San Diego.

Otherwise, in mid-March 2007, everything was unremarkable and routine for the *Sorcerer II* expedition. But not for the scientists at JCVI and in dozens of collaborating labs, institutes, and universities around the world. From Rockville to La Jolla, and to Halifax, Barro Colorado Island in Panama, the Galapagos, and Brisbane, Australia, a slew of scientists, the media, and the world at large were hearing about the publication of the *PLoS Biology* special issue, the "Global Ocean Sampling Collection."[1] The issue contained a wealth of information about the first half of *Sorcerer II*'s voyages from Halifax to the Galapagos, plus the Sargasso Sea—research papers, essays, reports, and editorials covering its first forty-one samples, taken roughly every two hundred miles just as *Challenger* had taken its samples in the 1870s, including in several exotic locales like the sulfur seep and the mangrove swamp in the Galapagos.

This was the first cache of scientific findings from the circumnavigation, a rich collection of data and analyses that would help launch thousands of additional studies and papers and

Pictured at the National Press Club on March 13, 2007, are Emma Hill (an associate editor of *PLoS Biology*), Victor Gallardo, Larry Smarr, Bob Friedman, and Craig Venter. Partially visible at the right edge of the photograph are the faces of Aaron Halpern and Doug Rusch.

PhD theses, and inspire other expeditions and projects seeking to understand microbes on a global scale.

Two of the biggest takeaways from the *PLoS Biology* collection were vindications of a couple of Craig Venter ideas about the microbial world: first, that it made scientific sense to think globally rather than locally; and second, that shotgun sequencing was a potent new tool to enable the identification of virtually any microbe or virus, and to boost the understanding of different genes and organisms and their characteristics and functions.

The *PLoS Biology* paper and other findings from the circumnavigation are discussed in detail in Chapter 7. For now, suffice it to say that the project was producing a tidal wave of new data about Earth's ocean microbes that scientists in 2007 were only just beginning to tap into, even as *Sorcerer II* kept going to collect more.

If the appearance of *Sorcerer II* in Loreto, Mexico, and the release of the *PLoS Biology* collection seemed like a scientific coda to the circumnavigation project, it wasn't. This was just the beginning of what would be another eleven years and tens of thousands more miles covered by *Sorcerer II*—and on other vessels, and on the land, and even in the sky, as Craig and the team combed the Earth for millions and millions more organisms.

∀

AFTER LORETO, *Sorcerer II* collected more samples from the Sea of Cortes, some of the thirty samples it would take in three months after leaving Annapolis. While sailing on this tranquil, sapphire-blue sea surrounded by desert, the team entertained guests that included Google founders Larry Page and Sergey Brin, who had joined *Sorcerer II* for its sail from La Paz on the Baja Peninsula to the island of Espiritu Santos. After Loreto, the ship continued its journey to San Diego, the new home of JCVI-West on the campus of UC San Diego and the Scripps Institution of Oceanography. *Sorcerer II* then took a breather for almost three months to be serviced and prepared for further explorations.

On June 25, *Sorcerer II* was ready to set out again, this time on a five and-a-half-month voyage to explore and sample waters up the Pacific coast to Puget Sound in Washington and onward to Alaska. The team stopped frequently to take samples at places like Green Harbor and Ketchikan in Canada, and in Glacier Bay

Sorcerer II approaching an ice field below Lamplugh Glacier, Alaska, July 2007.

Whales bubble-feeding next to *Sorcerer II* outside Glacier Bay Alaska, 2007.

National Park, about five hundred miles from Anchorage. Sidling up to melting glaciers at the park, they took two samples of water gushing from the thawing ice. *Sorcerer II* then turned around and headed back south to San Diego. It arrived on September 6, 2007, carrying eighteen more samples.

For the next eighteen months, Jeff Hoffman's team and other scientists continued to collect samples in the waters off San Diego and along the coasts of California, Oregon, and Washington. JCVI researchers also gathered samples with the Scripps Institution of Oceanography and the University of Washington, and collaborated with the Woods Hole Oceanographic Institution and other partners to collect deep samples onboard the research vessel *Atlantis* and its deep-ocean submersible, *Alvin*.

ON MARCH 19, 2009, Craig assembled a group of scientists and supporters on a dock in the San Diego harbor to announce that everything was ready for them to head off on another long, seemingly quixotic expedition he dubbed "*Sorcerer II* Expedition II," its official name being the Beyster and Life Technologies Research Voyage of the *Sorcerer II* Expedition, after its major funders: the San Diego scientist, business leader, and philanthropist J. Robert Beyster; and the sequencing company Life Technologies. This two-year voyage was setting out primarily to explore the three seas of Europe: the Mediterranean, Baltic, and Black. The team would get there by sailing back along the Pacific coast of Central America to the Panama Canal, and then east across the Caribbean and the Atlantic Ocean.

As always, the expedition was part of Craig's effort to sequence "the genes of the world," this time primarily in Eurasian seas largely isolated from the greater expanses of the world's oceans. Craig and the other scientists expected that they would find huge variations in microbial populations living in different ecosystems where the water varied significantly in terms of salinity, temperature, and nutrients, and because of human activity.

A specific objective of the expedition was to focus on collecting and charting populations of phytoplankton, microbes in the oceans that include cyanobacteria and certain species of algae like diatoms, photosynthesizing dinoflagellates, and blue-green algae. Phytoplankton live near the ocean surface, absorb carbon dioxide and sunlight, and convert these into over forty percent of the Earth's oxygen, leading some researchers to call the ocean the real "lungs of the planet."[2] It's a moniker usually given to tropical rain forests, which also absorb carbon and "exhale" oxygen. Phytoplankton at the bottom of the food chain also provide nourishment to everything from shrimp and fish to whales, and ultimately to humans.

In recent years, scientists have been warning that dumping carbon from burning fossil fuels and other pollutants into the atmosphere and the oceans is negatively affecting phytoplankton. It is therefore stressing the global marine microbial ecosystem that supports the production of all that oxygen and absorbs and sequesters about forty percent of the carbon that ends up in the atmosphere.[3]

To better track phytoplankton and other microbes, scientists at JCVI outfitted Expedition II with upgraded instruments to measure ocean conditions such as chlorophyll, salinity, pH, and oxygen down to a depth of one hundred meters. *Sorcerer II* scientists also had access to satellite images that tracked phytoplankton blooms from space, providing a color-coded roadmap for where to look. Other satellites monitored shifts in ocean currents and temperature fluctuations on a planetary scale.

"Yesterday afternoon, the irrepressible Venter himself was barefoot," reported the tech website *Xconomy* about *Sorcerer II's* launch event in San Diego on March 19, "as he hosted a dockside bon voyage party while an enormous white flag emblazoned with a blue insignia of discovery flapped lazily in the boat's rigging overhead."[4] Seeing off the crew was San Diego Mayor Jerry Sanders, JCVI's Ham Smith, and representatives from San Diego–based organizations and companies paying for the trip. These included private sponsor Bob Beyster and the CEO of Life Technologies, Greg Lucier. That morning, Lucier announced that his company was developing a new and ultra-fast technology called "single molecule sequencing," which the company hoped would drive down the cost of sequencing, say, a human genome to around $10,000. This was a bargain back in 2009 when full *Homo sapiens* genomes cost in the tens of thousands of dollars each to sequence. JCVI had agreed to try out the new technology, but apparently it didn't pan out.

Still, it's worth noting that in the background of the *Sorcerer II* expeditions, companies like Life Technologies and another San Diego–based company, Illumina, were rapidly driving down sequencing costs and creating smaller, nimbler technologies that by today have brought down the cost of a single human genome to around $200—while also reducing the time it takes to sequence a human genome from years to weeks, then to days, and now to hours.[5]

For months, Craig and a team from JCVI had been preparing for Expedition II with the usual flurry of organizing supplies, making repairs, and getting permission to sample from over fifteen countries. They also set up collaborations with dozens of scientists and thirteen research institutions, including the Universidad Nacional Autonoma de México, the Smithsonian Tropical Research Institute in Panama, the University of the Azores, the Plymouth Marine Laboratory in the UK, the Max

Planck Institute for Marine Microbiology, and Stockholm University, among others.

"The team set sail under sunny, clear skies and cool breezes despite reports of a possible offshore gale," reported the JCVI blog, which was now published online. Posts from this portion of the voyage were written by JCVI microbiologist Jeff McQuaid. He was temporarily taking over Hoffman's role as the maestro of sampling from the back end of *Sorcerer II,* while Hoffman worked on other projects.

"Within hours of heading out, we were sampling the waters off the Coronado Islands near the US / Mexican border and plotting our sampling schedule for the next few days," wrote McQuaid. This was the first stretch of water where the scientists would be collecting seawater rich in phytoplankton, guided by satellite images that picked up the telltale signatures of phytoplankton blooms in the ocean, particularly off rivers and cities where the waters are rich in nutrients like nitrogen from pollution and runoff.

Mostly, the satellite images flashing up on monitors in the ship's lab and main salon that afternoon were taken from SeaWiFS, NASA's global orbiter tracking chlorophyll in the ocean. Chlorophyll is the pigment found in phytoplankton (as well as plants) vital for photosynthesis. It's the green, inky color you see from a boat when chlorophyll-rich phytoplankton gather in dense blooms. From space, SeaWiFS sensors could detect the phytoplankton blooms in great detail, and provide data to computers at NASA, which in turn converted the data into images of blooms of different colors indicating the varying densities of chlorophyll and therefore all the phytoplankton in an area.

These color maps were then superimposed on territory maps, in this case of Baja and the west coast of Mexico, with purples and light blues showing up where algae blooms were minimal, out in the open sea, and greens, yellows, and reds showing up where the algae blooms got thicker, near coastal cities and areas of human activity. Some of these blooms were so substantial and persistent they had been named by scientists.

"The Ilsa Cedros bloom is halfway down the Baja peninsula on the west side," reported the *Sorcerer II* log, referring to an island on the monitor's map that was surrounded by blotches of red, indicating thick blooms of algae. Other mega-blooms showed up on the satellite images around Cabo Corrientes and just south of Puerto Vallarta. "Our satellite data indicates a large bloom extending 25 miles off the coast," the post continued, talking about a patch of sea off central and southern Mexico where *Sorcerer II* would be sailing over the next couple of days. "As we enter the bloom the water turns an intense green, and there are numerous fish feeding in the area."

A few hours after departing San Diego, *Sorcerer II* hit strong winds and rough seas that eventually caused too much roiling to take further samples. "Winds have picked up considerably in the last 36 hours," reported McQuaid in the log, "and tonight they are blowing in the 25 to 30 knot range, below gale force but still too strong to safely deploy our instrumentation. We sail past the plankton bloom near Cedros Island without stopping, but you can see the sparkle of the bioluminescent plankton as the *Sorcerer* plows on through the night."

By March 26, the wind had died down. "Sampling today starts before sunrise when we arrive at Puerto Vallarta," wrote McQuaid. "In conjunction with our Mexican collaborators, we are investigating the influence of coastal development, particularly intensive tourism, on marine microbiota, so we take a sample of surface water in Banderas Bay and leave the harbor with the rising sun."

Later that day, the team took a sample south of Puerto Vallarta in the middle of a bloom twenty-five miles long. "As we enter the bloom the water again turns an intense green," wrote McQuaid on JCVI's blog, "and there are numerous fish feeding in the area. Sampling conditions are ideal: bright sunshine, light winds, moderate swell. We deploy a large plankton net which rapidly fills with algae and zooplankton." The latter term refers to microbes that are eukaryotes, with the same basic cell types as humans.

"From the aft cockpit we deploy a CTD equipped with a sampling hose," continued McQuaid, explaining that "a standard CTD measures conductivity, temperature and depth: our unit also contained a pH probe and a fluorometer for measuring chlorophyll concentration." The device was a new addition since the global expedition. "As we lower the CTD through the water column, we generate a profile of the ocean at Cabo Corrientes down to 40 meters in depth."

The instruments revealed a layer of water at about eight meters where phytoplankton were thickest and oxygen levels highest. The scientists hoped that sampling across these layers would allow them to sequence undescribed microorganisms that might help to explain the relationship between photosynthesis and respiration in the ocean, and how exactly the "lungs" of the planet work.

The voyage continued down the isthmus to Acapulco and then to the Gulf of Tehuantepec in southern Mexico before reaching Guatemala, El Salvador, Honduras, and finally Nicaragua. There, the ship was buffeted by the powerful winds that blow that time of year across the Central American isthmus and contribute to an upwelling of nutrients in the Pacific coastal waters. "These nutrients enable phytoplankton to grow," wrote McQuaid, "and as we approach the southern end of Nicaragua, the water again takes on a greenish hue, and we note large amounts of sea turtles on the surface of the water." As the ship approached Panama, the chlorophyll was so thick that it coated the lines and hoses with a green layer of algae.

Sorcerer II continued south to Panama City, entering the Panama Canal on April 6 and arriving back at the Smithsonian Tropical Research Institute on Gatun Lake that afternoon. The team stopped to take follow-up samples to compare with those collected during the global expedition, even though recent dredging and widening of the canal and lake had stirred up a tremendous amount of silt in the water. The filters were instantly clogged, and what they took away looked more like a soil sample than a water sample.

Three days later, *Sorcerer II* exited the canal. "We are now out in the warm and saline Caribbean Sea," said McQuaid, "and the waters are blue because there is very little algae in them. We dropped the CTD and barely got 0.25 micrograms of chlorophyll per liter all the way to the 50-meter mark," which wasn't much. A few days later the ship arrived in Fort Lauderdale, where the team shipped home their frozen samples. "This week we prepare to depart for Bermuda and the Azores and will continue on to Plymouth Marine Laboratory in England," wrote McQuaid.

While *Sorcerer II* was in Florida, Hoffman returned to take command again of the sampling, after a break from life aboard the ship of almost two years. Recently, he had traveled to take samples in Antarctica, one of six trips between 2006 and 2015 to the bottom of the planet taken by JCVI teams sans *Sorcerer II.* (We'll describe one of these later in this chapter.) Hoffman also resumed log-writing, a task that for this trip became more formal and less "inside baseball" (and perhaps less entertaining) than during the circumnavigation. With the logs now being posted on JCVI's public website, Hoffman, for instance, no longer referred to "Craig" but instead called him "Dr. Venter."

The expedition was also joined in Florida by Karolina Ininbergs, a researcher at Stockholm University and an expert on marine cyanobacteria and nitrogen fixation. Ininbergs was part of a research group at Stockholm University run by Birgitta Bergman,

Professor Birgitta Bergman and her associate Karolina Ininbergs, a researcher, both of Stockholm University's Department of Botany.

a professor in its Botany Department. Two Swedish foundations, Baltic Sea 2020 and Stiftelsen Olle Engkvist Byggmästare, also helped support the expedition in collaboration with Berman's lab.

"The earliest life forms on this planet are thought to be early ancestors of cyanobacteria," Ininbergs wrote in the ship's log, referring to the tiny bacterial phytoplankton that are prolific all over the planet and critical to sequestering carbon from the atmosphere and producing oxygen. "They were the first organisms capable of photosynthesis." In her research she was working to better understand the cyanobacteria in the Baltic Sea, how they affected oceans and the planet, and how they had evolved over the last two to three billion years.

ON TUESDAY, April 21, *Sorcerer II* departed Fort Lauderdale. Buffeted by thrashing winds and seas all the way to Bermuda, it arrived at this island a thousand miles away four days later at 5:00 PM. Two samples were collected there, including in the Sargasso Sea, and the CTD data confirmed what the JCVI science team had expected in terms of the presence of chlorophyll in the water—that there wasn't much green stuff in this deep, nutrient-poor stretch of open water.

In Bermuda, the crew was surprised to find that the permits for sampling had not yet been approved by the government, despite Craig and JCVI's close ties with the Bermuda Institute of Ocean Sciences and its director, Tony Knap. Disappointed, they waited for three days to no avail before setting sail for Europe. They quickly caught a fast wind and were about twenty miles off the coast—near Hydrostation "S," where *Sorcerer II* had taken samples in 2003—when the harbormaster from St. George's called on the VHS radio to inform them that the permits had come through. The ship was still close enough to the harbor to turn around, and they were able to bring aboard a single sample in Bermudian territorial waters.

Setting off once again into the open sea, the team collected eleven samples on the two-thousand-mile jaunt to the Azores, their first stop before reaching the European mainland. "On the North Atlantic transit weather played a key role on when and where we could sample," wrote Hoffman. "A few days out from Bermuda we were informed that a weather system was coming from the north, and we had to move south to avoid the brunt of it." For most of the crossing the sea was a combination of jostling waters and light winds, which meant the sail wasn't very useful in stabilizing the ship. "We just rolled from side to side," Hoffman reported, adding that "it's tough to walk, work, eat, or sleep."

Pausing in Horta, a small town on the Azores' Faial Island, the crew collected samples in collaboration with microbiologist Sergio Stefanni from the Department of Oceanography and Fisheries at the University of Azores. This research center, situated in the middle of the deeper parts of the Atlantic, appropriately studies deep-sea ecology, fisheries, and conservation with a special interest in seamounts and hydrothermal vents. "Stefanni helped with choosing sites that included the unusually warm waters, and hydrothermal sulfur vents," reported the log.

Sorcerer II sailed northeast from the Azores into an intense storm and was forced off course almost to Spain to avoid the worst of it. "During this time, we experienced winds up to 50 knots and seas ranging from 15–20 feet," wrote Karolina Ininbergs. "As you could imagine the weather put a halt to the sampling and the crew focused on the weather and making it to England safe and sound."

They arrived in Plymouth on Monday, May 18, two days later than expected thanks to the battering weather, but still in one piece. "Plymouth is a significant location for us since Charles Darwin embarked aboard HMS *Beagle* from this same site 178 years ago," wrote Ininbergs. Waiting for them on the dock were two microbiologists mentioned in earlier chapters: Chris Dupont from JCVI, and Jack Gilbert, then a senior scientist at the Plymouth Marine Laboratory and later a professor at UC San Diego.

Joined by Gilbert and Dupont, the team took the first British sample off Plymouth on May 22. "We had heard rumours about blooms of *Phaeocystis,* a conspicuous bloom-former in the North Sea and English Channel," wrote Ininbergs. "When it blooms, it turns the water reddish-brown in colour." (Since the arrival of Ininbergs, British spellings had begun appearing in the log.) Foam also can be frothed up by waves and wind when gelatinous globs of *Phaeocystis* die and degrade, creating a sudsy surface brew that many people mistake for pollution.

Sorcerer II followed the research vessel *Plymouth Quest* out to Plymouth Marine Laboratory's continuous sampling stations, where the team recorded changes in temperature, oxygen, and pH at depths ranging from the surface down to thirty-five meters. The instruments also showed a gradual increase in chlorophyll as the depth increased, which Ininbergs said was due to photoadaptation—the evolution of deep water plankton that produce more light-absorbing pigments because there is less light.

"We were delighted to find our filters full of microorganisms after filtering 200 L of seawater from the two depths," reported the log. "Upon opening the filter casings, we were hit with a very tangy sulfidic aroma caused by dimethyl sulfide (DMS for short). This gas, which is literally the 'smell of the sea,' is the result of marine plankton degrading dimethylsulfoniopropionate (shortened to DMSP for sanity). Most phytoplankton produce lots of DMSP, so you essentially have a steady supply of DMSP to bacteria. Some bacteria want the sulphur so they metabolize the DMSP. Others degrade it to DMS, which gives you that peculiar smell. A curious side effect of DMS is that when in the atmosphere it acts as cloud condensation nuclei. In simpler terms, lots of DMS production means lots of clouds, which reflect the sun's energy away from Earth. Therefore, in direct contrast to the carbon dioxide, DMS is a 'global cooling gas.'"

Sorcerer II spent another week in the UK, visiting labs and research facilities, before departing on June 3 to sail up the English Channel and across the North Sea to Germany. "On June 6th we arrived on Helgoland," wrote Ininbergs, "an island about 70 kilometers from the mainland." The team collected a sample from the Long-Term Ecological Research Site, a buoy located just outside the main harbor of Helgoland in the southeastern corner of the North Sea. A team of German scientists led by Frank Oliver Glöckner, head of the Microbial Genomics Group at the Max Planck Institute for Marine Microbiology, came along. A German account of their visit described these local scientists coming on board *Sorcerer II* "like a swarm of grasshoppers."[6] At least that's how the German was translated into English (by the Germans).

"In former times a buoy was anchored at this site to hold a cable connecting the main island with the Dune, a sandy smaller island populated mainly by seals and tourists in the summer months," continued the German team's account. It had yielded samples for almost fifty years allowing for monitoring of "food

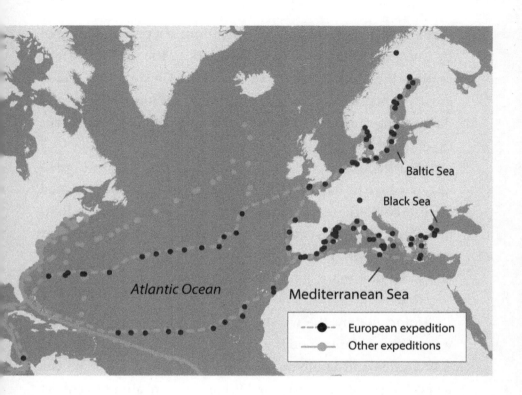

Baltic Sea

Black Sea

Atlantic Ocean   Mediterranean Sea

- - -●- - - European expedition
———●— Other expeditions

web interactions and the influence of climate change and the diversity of microbial communities in the North Sea." The Germans planned to use the samples and data from the *Sorcerer II* sampling in an effort called MIMAS (Microbial Interactions in Marine Systems), which "generates and integrates diversity, metagenomic, metatranscriptomic and metaproteomic data with contextual data like temperature and nutrient concentrations." The final day of this collaboration ended with an Anglo-American-German barbecue.

⋎

THE NEXT DAY, *Sorcerer II* transited the Kiel Canal. This narrow, sixty-one-mile waterway, the construction of which was completed in 1895, cuts across the bottom of a peninsula shared by Germany in the south and Denmark in the north. By transecting the German state of Schleswig-Holstein, the canal greatly expedites what would otherwise be a journey of several days sailing north into the Baltic Sea and around the northern tip of the peninsula.

From there, *Sorcerer II* went on to pick up Craig in Copenhagen, where he had flown to rejoin the expedition as it headed across the Baltic to Sweden—"my home," wrote Karolina Ininbergs in a blog post, "and one of the main destinations of our 2009 expedition. It was a proud and special moment for me when the first mate, John [Henke], hoisted the Swedish courtesy flag."

"Unfortunately, the weather has not been cooperating and was putting a damper on the excitement," Ininbergs continued. "My friends and family in Stockholm tell me it has been the worst June weather in 50 years! When we were about to collect our first sample in the Baltic in winds up to 30 knots, rain and cold, Jeff Hoffman felt the need to pull out his thermal underwear that he uses in Antarctica. For some reason he didn't seem overjoyed when I screamed through the wind, 'Welcome to the Baltic!'"

"With Dr. Venter at the helm, and in spite of the weather, we made our way north along the Swedish coast," wrote Ininbergs, "and after a brief overnight stop in the island of Öland we reached our first Swedish sampling site. Because of the cold weather I didn't expect to find much of my beloved cyanobacteria, but the CTD cast revealed a chlorophyll max at around 15 meters and just by looking at the 3.0 μm filters we could see the spiky colonies of *Aphanizomenon*, one of the common bloom-forming cyanobacteria in the Baltic Sea. We also saw some of the toxic dinoflagellates *Dinophysis*, and it will certainly be interesting to see what kind of smaller bacteria and viruses are associated with these phytoplankton communities."

After spending the night on a nearby island, on June 15 the team headed into the channel that winds through the islands of the Stockholm archipelago and leads to the capital city. They happened to arrive when boats competing in the famed Volvo Ocean Race—formerly known as the Whitbread "Around the World Race"—were passing through. The race had begun in Alicante, Spain on October 11, 2008, and would end a few days later, on June 27 in St. Petersburg. Craig joined Erling Norrby, who lives in Stockholm, to sail on one of the competing boats. Named *Il Mostro*, it was sponsored by PUMA, the shoe and clothing company. "Craig and I had the privilege of joining one of the boats to watch the exceedingly well coordinated teamwork when operating the fast boats, for example when tacking," said Norrby. "They turned over the helm to Craig for the final leg into Stockholm."

"*Sorcerer II*, along with hundreds of other boats, watched the start of the race and then for the next three hours followed the boats into Stockholm," remembered Hoffman. *Sorcerer II* docked in the Volvo Race Village with the competing boats, and the crew attended a reception for the racers at the Royal Palace.

During the next week, Craig gave lectures in Stockholm and welcomed King Carl XVI Gustaf on board *Sorcerer II*. According to Craig, the two of them drank half a bottle of Jameson whiskey, "just the two of us. He was supposed to stay for ten minutes, but instead stayed for a couple hours." Craig also spoke to a reporter from *Xconomy* about the expedition in the Baltic Sea, taking

Craig Venter hosting His Majesty Carl XVI Gustaf, King of Sweden, on board *Sorcerer II* in Stockholm, Sweden, June 2009. Behind Craig from left to right: Jeff Hoffman, Karolina Ininbergs, Jeremy Niles. Behind them: Heather Kowalski. To Craig's left is King Carl XVI Gustaf and his team.

issue with the long-held assumption that the Baltic Sea is a brackish body of water with lower diversity of microbial life than the salty oceans. "That is the conventional wisdom," said Craig. "But a hallmark of my career has been to not take the conventional wisdom for granted. The Baltic is a body of water unlike any other, with tremendous salinity and temperature gradients and very influenced by human activity."[7]

"The morning of June 25th we left Stockholm and followed the Volvo race boats into the Baltic to watch the start of the last leg of the race to St. Petersburg," wrote Hoffman in the ship's log. From there, Craig and Charlie Howard steered through some tricky waters along the Stockholm archipelago, assisted by avid sailor Erling Norrby, who knew this route well. The ship was headed north to the Gulf of Bothnia, which runs between Sweden and Finland.

"The last days of fantastic summer weather had warmed the water to about 20° C in the little bay," wrote Ininbergs, "and we found our first specimen of *Nodularia spumigena,* the most conspicuous toxin-producing, bloom-forming cyanobacteria in the Baltic Sea. Just like *Aphanizomenon* this cyanobacterium has heterocysts, cells specialized for fixing nitrogen from the atmosphere." The ship kept going through the night, which was never entirely dark in this near-arctic summer. The sun rose at 3:30 AM. "The following afternoon we reached our next sampling site," wrote Ininbergs, which was "one of the Helsinki Commission's monitoring stations in the Baltic Sea." For thirty-five years, these stations had been monitoring the effects of pollution and human activity in the Baltic Sea.

On July 4, the team arrived in Luleå, a large city along the Swedish northern coast of the Gulf of Bothnia whose port is used for shipping iron for the mining industry. Here, the team recorded the lowest salinity yet on the voyage. In the standard measurement of practical salinity units (psu), it was only 2.9 psu at the surface at a temperature of 10 degrees Celsius.

Sorcerer II docked in Luleå for a few days while Hoffman and Karolina Ininbergs prepared to lead a road trip 130 miles above the arctic circle to take samples at Lake Torneträsk, a major freshwater feeder into the Baltic Sea. They departed Luleå on July 6, in a rental car filled to the brim with equipment, and soon were cruising through a remote land of snowcapped mountains, thick green forests, and road signs warning of moose crossings. The next day, they reached the lake and walked to the end of a sixty-foot pier to fill up a fifty-liter cannister with water. "Lake

Karolina Ininbergs and Jeff Hoffman filtering water samples 130 miles north of the Arctic Circle, July 2009. Samples were taken from Lake Torneträsk, a major freshwater feeder into the Baltic Sea, Abisko, Sweden.

Torneträsk doesn't thaw until June," reported Hoffman, "so the water we collected by hand was 7° Celsius, 44° Fahrenheit. The water was crystal clear so the team thought they would have to suck in huge amounts of water to get enough biomass for DNA sequencing. But after only 200 liters, all three sizes of filters were mashed full of microorganisms and had distinct colors." This was far more life on the filters than they usually saw in open ocean samples when a total of four hundred liters was filtered.

On July 9, Hoffman and Ininbergs returned to *Sorcerer II*, which next sailed south along the east coast of Sweden. After another brief stop in Stockholm, the ship kept heading south, taking samples and pausing briefly above the Landsort Deep, the deepest part of the Baltic Sea at 459 meters. Arriving late in the evening, the scientists lowered their instruments down, down, down as they checked oxygen levels, finally reaching zero oxygen just below seventy meters—the depth where they collected their sample.

Throughout the expedition in the Baltic, Karolina Ininbergs had been looking for a cyanobacterial bloom—normally a summer occurrence—with no luck. Even satellite images weren't helpful because their sensors in space were being blocked by cloudy skies. Then, on July 18, they sailed right into a bloom. "Well . . . almost a bloom," wrote Ininbergs. "We spotted a lighter streak in the water and when taking a closer look, we could see that the water was filled with small whitish aggregates of some sort. In the microscope we could confirm that these aggregates were indeed cyanobacteria, and that *Nodularia spumigena* seemed to be dominating the sample. Due to the whitish colour we suspect that the filaments were dying since these types of blooms are typically more yellow."

On July 31, the expedition began a slow trip around the south end of the Scandinavian peninsula, from Sweden to Norway, where Ininbergs debarked, planning to rejoin the expedition the next summer in the Mediterranean. From Oslo, *Sorcerer II* headed south, making its way through the North Sea and down the Atlantic coast of mainland Europe, turning east in southern Spain and sailing through the Strait of Gibraltar into the Mediterranean Sea. On September 21, it arrived in Valencia, Spain. This was to be *Sorcerer II*'s home for the next seven months as the ship was refurbished over the winter for the Mediterranean cruise launching in the spring of

2010. In the interim, the scientists and crew would take a break and head home.

Everyone except for Jeff Hoffman. After helping secure *Sorcerer II* in the dry dock in Valencia, he did not hop on a plane back to his home in Rockville, Maryland. Instead, Hoffman flew on November 9 to yet another exotic exploration, one that was almost ten thousand miles away, and way south of Valencia.

His destination: Antarctica.

y

HOFFMAN HAD BEEN to this icy continent four times before to collect samples, so he knew something about the extreme conditions there. Yet his previous visit hadn't prepared him for what walloped him and his team of scientists soon after they arrived and headed out to take samples on the Ross Ice Shelf. A "Condition One" Antarctic storm swept in just a few hours after they left McMurdo Station, the sprawling town where scientists converge in the continent's summer to conduct research. In a flash, winds whipped up to sixty-three miles per hour, and wind chill made the temperature feel like minus-100 degrees Fahrenheit. Visibility was less than a quarter-mile.[8]

The group, including collaborators from JCVI, University of Southern California, and Woods Hole Oceanographic Institute, had come to Antarctica on a three-week mission to investigate phytoplankton in the Ross Sea. Like the other JCVI expeditions to Antarctica in this era, the effort was mostly to see what was there and to use tools like metagenomics honed during the *Sorcerer II* expeditions to characterize the phytoplankton they found.

Andrew Allen, a microbiologist and oceanographer at JCVI and Scripps Institution of Oceanography, took the lead in much of this work on the field trips down under. "We were able to piggyback on the momentum and energy of GOS," he said, referring to the Global Ocean Sampling expeditions on *Sorcerer II*. "In those days we were mostly studying metagenomic topics and major phytoplankton and taxa in Antarctica." In later expeditions, Allen and other scientists turned to studying the impact of temperature increases in the Southern Ocean and how climate change was affecting phytoplankton populations and the patterns of nutrients that start in the Southern Ocean and flow into the southern Pacific Ocean to help feed plankton and fish there.

This expedition was recorded in meticulous detail in blogs posted to the JCVI website.[9] The entries give a flavor of some of the JCVI sampling that was done off *Sorcerer II*—in this case, way off.

The scientists started with satellite images that showed patches of phytoplankton appearing in open patches of water on the shelf, where rising temperatures from global warming had exposed parts of the sea to sunlight for the first time, in some cases, in millions of years.

On the morning of November 15, when the expedition left McMurdo Station, the sky was a crisp blue. The forecast said snow, but nothing hinted at a "Condition One" storm as the team headed out onto the shelf covered with intensely white snow. Two large snowcat-style vehicles, one red and one yellow, hauled hundreds of pounds of gear, including pumps, filters, drills, chemicals, and other equipment they needed for sampling below the ice, plus two snowmobiles, several drums of diesel, tents, camping gear, and a generator.

Other JCVI scientists had visited Antarctica on expeditions in 2006, in late 2007, in early 2008, and earlier in 2009. They

Jeff Hoffman filtering samples from Ace Lake Antarctica, December 2006.

would visit way down under twice more after the current trip, in 2013 and in 2015. The mission for all of these trips was to collect microbes inside glaciers, in frozen lakes, under layers of ice and snow, and in the holes that form in ice shelves as they melt (or are opened up by powerful winds). Called "polynyas," these watery patches act like windows to let in sunlight and ignite life in the sea below the exposed surface—plants, fish, algae, and other microbes. *Polynya* is a term borrowed from the Russian word for "little field."[10]

"Antarctica and the Southern Ocean are keys to what's happening globally with climate change," said Craig, years later. "We wanted to check on changes in micronutrient availability, temperature, and oxygen, and how this impacts phytoplankton, and things like growth rates and nutrients and community composi-

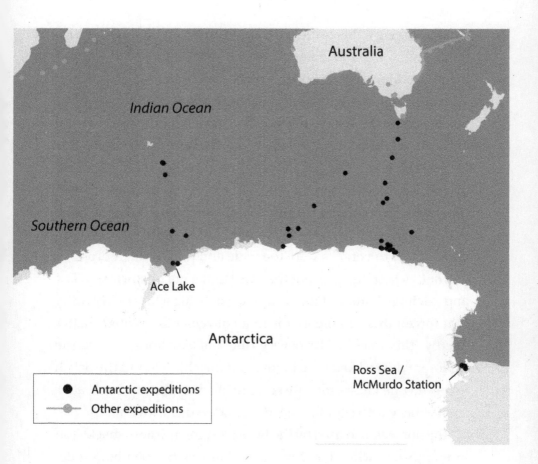

tion. Ongoing climate change is affecting the polar regions faster than any other place on the planet, so there was urgency to catalog what was there and how it was changing, to establish a baseline of fundamental scientific knowledge."

"Initially we followed the Cape Evans sea-ice traverse out of town," wrote Jeff McQuaid, the microbiologist who had headed up sampling on *Sorcerer II* earlier that year, during the first leg of Expedition II from San Diego to Florida. He was now accompanying this Antarctic mission and oversaw the writing of the expedition log. "This route is a flagged and reconnoitered sea ice 'highway' so we were able to travel at maximum speed without having to stop and check the thickness of the ice. (Note that 'maximum speed' in a Sno-Cat is about 3 mph.) After two and a half hours on the sea-ice highway, we turned off the traverse and cut fresh tracks across the snow toward Inaccessible Island. The island is a big hunk of basalt rising straight out of the Ross Sea, and as we rounded the tip of the island, we saw a line of Weddell seals stretching off into the distance, evidence of large cracks in the ice. We tested several possible routes through the area, but all the cracks were too wide and thin to safely cross."

That's when the team got the word that a major storm was fast approaching. Minutes later it struck, a howling white monstrosity that forced them to hightail it to a conveniently located shelter nearby. This was the "Discovery Hut" built by Robert Scott, the explorer who trekked here in 1912 and may have found the South Pole—no one knows for sure since he died on the way back, after a relief party with supplies failed to rendezvous with him.

The hut was also used by the British explorer Ernest Shackleton in 1915 to 1917, when his icebreaker ship was trapped in and destroyed by winter ice nearby. The structure allowed his stranded men to survive until a team he led managed to make it back to civilization and brought help. Over the years, the hut had been abandoned, reused, and finally updated and renamed "Hut 12" in the late twentieth century. "We had good radio communications with McMurdo Station," wrote McQuaid as the wind howled outside, "so we were never in danger, but we were certainly not going anywhere anytime soon. Hut 12 was our new home."

Hours went by as the storm raged and "morning" arrived, although the concept of morning was relative in a place where,

at this time of year, the sun shines twenty-four hours a day—
and, in this case, when the sky was blotted out by snow that showed
no signs of abating. The team remained trapped in Hut 12: "The
wind was so strong that snow began drifting up through the dive
hole in the warming hut, and the windows completely glazed
over with snow. Needless to say, trips outside were kept to a
minimum, and we were sure to keep hot chocolate and coffee
at hand."

The team hated to be stuck indoors, but there was still plenty
of science to do. "We considered it an incredible stroke of good
luck to be stuck in a hut with a convenient access hole to the
Ross Sea," wrote McQuaid, referring to an opening cut in the ice
that formed part of the hut's floor. "Using this hole while staying
warm inside the shelter, the team was able to collect plankton
from the sea ice and drop a sediment trap below the ice to
gather dead or dying plankton that sink to the bottom of the sea
and can help scientists understand why they might be dying.
The team was also astounded when a Weddell seal popped up
through the ice hole. Throughout the day Weddell seals would
use the hole in the ice as an access point for air."

"Eventually the storm slowed down enough that we could
start to dig ourselves out," he continued. "There was still snow
blowing down off the Erebus glaciers, but by our third day in the
hut, visibility started to improve enough that we could shovel a
path and clear some of the vehicles to use to get back to McMurdo."
The Pistenbully, their largest vehicle, "needed to be warmed for
two hours before we could start it up, but once it was running,
we decided to head back to McMurdo."

It took another day for the storm to blow itself out, but by
November 18 "the wind and driving snow had abated, and we

drove our Pistenbully back out to our temporary shelter [Hut 12]. It took several hours of digging to clear the snow away from our other vehicles," the ones they had left behind, "but once we started driving away from the hut we quickly ran into another problem: the snow was so deep that our sleds and vehicles would bog down in the snowdrifts. Often, we would pull out the Pistenbully, only to have the Sno-Cat get stuck and have to dig that out." An entire day went by digging vehicles out of the snow, and in the process, the team broke the sled hitch and winch. Eventually they were able to drag their research sled to a suitable spot for taking a sample of plankton, and they set up camp.

The team spent the next day setting up and drilling holes into the ice and collecting samples. "The sea ice is remarkably consistent until you approach the bottom layer," wrote McQuaid, "where diatoms and other phytoplankton have entrained themselves in the ice. By keeping to the underside of the ice, microscopic plankton are the first organisms to intercept light as it enters the ocean, but the plankton are still close enough to the unfrozen seawater to obtain dissolved nutrients and minerals. Living in the ice also protects the phytoplankton from grazing zooplankton like Antarctic krill." The team then ran the water they collected through the usual system of three sizes of filters, "which can take us most of the day."

They worked into the night, and by midnight were ready to wrap up and head back to McMurdo Station. "Mother nature had a different idea though," reported McQuaid. "As we were leaving, the winds picked up, and began gusting over 50 knots, while the blowing snow caused the visibility to drop. McMurdo operations informed us that the weather had just deteriorated again, and was Condition 1, meaning no travel whatsoever. We were pinned down by yet another Antarctic storm!"

This time, with no hut to hunker down in, the team slept on the floor of their research sled and in the back of the Pistenbully. "The wind shook and buffeted the vehicle all night, and at times the Pistenbully made this vibrating sound like we were just about to take off." By 6:00 PM the next day, the wind had died down enough for them to head back to McMurdo, leaving their equipment to dig out the next day.

Next up for sampling was a polynya that the team had learned about from satellite images and from planes flying low over the Ross Ice Shelf. On November 25, they set up their equipment at the polynya and began taking samples. They were interested to see how plankton in the open polynya were different from the phytoplankton they had isolated from areas locked in sea ice.

"We set up our plankton filtration station, and as we worked our audience of flightless birds grew," wrote McQuaid. "Emperor penguins are extremely curious animals, and they fearlessly let us know that we were in their domain. At times they would waddle around our filters and pumps, silently looking over our setup and seemingly unperturbed by our noisy gasoline-powered air compressor. Other times they would just stand around and preen." The team collected three separate samples of plankton. The first was for DNA shotgun sequencing, which JCVI scientists would use to identify the total complement of genes present in the seawater. Another sample was for mRNA, "the total transcriptome of the plankton," which would tell them what genes were actively being used by the microbes when they were collected. A third sample was slated for analysis of the microbe's proteome.

"It takes us about seven hours to filter sufficient volumes of water and harvest enough plankton for these analyses," McQuaid explained, "so in the meantime we sat out on beach chairs, kept an eye out to make sure our hoses and pumps didn't freeze, and watched the penguin parade."

The parade ended when the team was finishing the sampling and the penguins sensed they were about to leave. Queuing up on the ice in an orderly line, "they tobogganed back off into the ocean," wrote McQuaid. Soon after, the scientists headed back to McMurdo. "We were all tired and looked forward to some fresh food and heated rooms back in McMurdo Station."

In the years following this and other Antarctic expeditions, the scientists involved wrote and published dozens of studies. Examples include a 2010 paper in *ISME Journal,* coauthored by JCVI scientists and German-Australian microbiologist Torsten Thomas, along with other collaborators from the University of New South Wales in Australia. They offered up a first-time analysis of the green sulfur bacteria called *Chlorobiaceae,* which are important in Antarctica and elsewhere to how carbon and sulfur cycle globally in natural environments.[11] Another paper was published in *Molecular Phylogenetic Evolution* in 2014, reporting on a study of the mechanics of mutations in red algae that appeared in one of polynyas.[12] In 2019, a study in *Nature Communications* analyzed the role of iron and manganese in the survival of phytoplankton in the Southern Ocean.[13]

⅄

WHEN WE LAST heard from *Sorcerer II,* the vessel was being repainted, repaired, and retrofitted in Valencia. In early May 2010, the crew returned to Spain, including Jeff Hoffman, who had finally been able to spend some time at home in San Diego after Antarctica. Holing up in this ancient Iberian city about halfway up Spain's Mediterranean coast from Gibraltar, Charlie Howard and crew made final preparations for the next leg of Expedition II, which was set to travel east across the Mediterranean to Greece and Turkey, then onward through the Bosporus into the Black Sea, and finally back again to the United States.

Meanwhile, back in San Diego, Craig was about to make another big announcement involving the world of the very small: his seminal achievement of synthesizing a bacterial genome assembled out of As, Cs, Gs, and Ts in the laboratory and then inserting it into a cell and booting it up. David wrote about this achievement in the May 27, 2010, edition of *Fortune*.[14]

> Last week, Venter and a team at the J. Craig Venter Institute in San Diego . . . revealed to the world a microbe brought to life by DNA they had painstakingly assembled in their lab. Venter described it in a press conference last week as "the first self-replicating species we've had on the planet whose parent is a computer."
>
> The bacteria . . . couldn't do much more than survive and reproduce. But the proof of concept was there for a synthetic bug that Venter says will one day be programmable to churn out new drugs, bioengineered fuels, and vaccines, among other things.

The announcement was greeted with a combination of amazement at the scientific feat and alarm at the ethical implications of humans designing and creating life in a lab. Craig stressed that this technology would be used to benefit humans and the planet. Yet plenty still insisted he was playing God. The

Obama Administration was concerned enough that they somewhat hastily announced an order from the president that his
Commission for the Study of Bioethical Issues must prepare a
report on the ethics of synthetic life.[15]

The new organism was adapted from a natural bacterium,
Mycoplasma mycoides, that causes mastitis in goats, although
the genome was entirely constructed in the lab.[16] "The single-
celled organism has four 'watermarks' written into its DNA to
identify it as synthetic and help trace its descendants back
to their creator, should they go astray," reported the *Guardian.*[17]
"We were ecstatic when the cells booted up with all the watermarks in place," Craig told the UK newspaper. "It's a living species now, part of our planet's inventory of life."

This project was part of a continuum of experiments that
began with Craig's sequencing of *Haemophilus* in 1995 and
later led to the sequencing of bacteria collected on *Sorcerer II,* in
Antarctica, and beyond. "Building a synthetic genome for a bacterium in the lab helps us understand what we're finding on
Sorcerer in nature," said Craig, "including a lot of unexpected
and exotic stuff."

On May 17, 2010, as kudos and controversy erupted from the
announcement, *Sorcerer II* departed Valencia and began a mostly
leisurely sail, collecting fourteen samples along the Spanish,
French, and Italian coastlines. It made stops along the way in
Barcelona, Mallorca, Ille Maree, Capri, Rome, and other cities and
towns in Italy, finally reaching northwestern Greece at Atokos Island on August 1. *Sorcerer II* rounded the Peloponnesian peninsula and then island-hopped in the Aegean Sea, as the crew and
Bob Friedman in the States worked to secure permits and nego-
tiate agreements—among them, a stipulation from Athens that

a Greek monitor needed to be on board whenever they sampled. The delays this introduced were on top of delays already caused by a series of crazy, sometimes violent storms fulminating up and down the Aegean Sea all month. Yet more permits were required for coming and going in some areas due to the late-summer proliferation of tourists and boats.

On August 29, the ship arrived in Turkey at the ancient city of Çanakkale, southwest of Istanbul. "It would have been an uneventful one-night stop if it wasn't for the Byron Hellespont Bicentenary Swim," wrote Hoffman in the ship's log. "This yearly race allows you to swim the 3 miles from Europe to Asia across the Dardanelles. I would like to say that it was a great day for a swim, but that would be a lie," he added, noting that he participated in the race, having been a competitive swimmer in high school. "The winds were blowing between 20–30 knots," he reported, "very choppy seas and a 4-knot current. The 4-knot current was the biggest problem, it pushed the majority of the participants way past the finish line, including myself. We made it to Asia, but about a quarter of a mile downstream of the finish line."

Two days later, *Sorcerer II* arrived in Istanbul, sliding into the ancient harbor amidst every sort of ship imaginable. From the water, the crew could see the great dome of the Hagia Sophia, the sixth-century Byzantine cathedral turned mosque, and the domes and minarets of other mosques in a city brimming with buildings, markets, baths, and palaces, some in ruins, along with endless, twisting vibrant streets and alleys dating back to three great empires: Rome, Byzantium, and Ottoman. After a brief stay and meetings with Turkish collaborating scientists, *Sorcerer II* headed into the Bosporus, the narrow strait of water dividing Europe from Asia that is the entrance to the Black Sea.

"This transect is very important because you have huge amounts of water moving back and forth from the Black Sea and Aegean Sea," wrote Hoffman. "There are also big salinity gradients, not only in the surface waters from each body of water, but also in the water column due to the movement of water from each area. The salinity of the Sea of Marmara [south of the Bosporus] is slightly greater than that of the Black Sea but only about two-thirds that of most oceans."

Once in the Black Sea, the expedition sailed northwest to Nessebar, on Bulgaria's western shore—an ancient city that has been part of empires and kingdoms for three millennia. It's divided into two parts separated by a narrow, man-made isthmus, with the ancient part of the city on the peninsula and a more modern section on the mainland. *Sorcerer II* was headed there at the invitation of a well-known astronomer, Dimitar Sasselov, the director of the Origins of Life Initiative at Harvard, known for its work of identifying Earth-like planets. Craig had met Sasselov when he was a visiting professor at Harvard. The plan was for *Sorcerer II* to stay in the harbor for a day and a night while a lecture engagement took Craig, Heather, Craig's son, and Sasselov overland to the Bulgarian capital of Sophia.

While in Nessebar, the crew was "worked over by the Bulgarian mob," said Craig. "We arrived at night, and someone put fishing nets out to entangle boats as they entered the port. We had to cut them. So, when we docked, this brand-new Mercedes pulled up near our boat and two characters stepped out and they threatened us. They said we'd ruined their fishing nets. They demanded $10,000 cash or, they said, there would be trouble. The captains of the other vessels in the harbor urged me to turn them down because they had all been subjects of extortion attempts. So these guys came back the next day and I said, 'hell no' to the money. They made threats that seemed potentially serious."

The next day, Craig informed Professor Sasselov of the shakedown attempt, and Sasselov called the attorney general of Bulgaria, whom he knew. "Bulgaria was attempting at the time to become part of the EU," recalled Craig, "but they were told that they had to clean up their mob problem. So, the last thing the government wanted was news of an American research vessel being held up by the mob." When the government sent armed soldiers to protect the ship, the gangsters left them alone while Craig delivered his lecture, and until they departed Nessebar the next day.

From Bulgaria, *Sorcerer II* headed back south through the southern Black Sea and the Bosporus and into the Aegean. For two weeks the team sailed and took samples in Greek and Turkish waters as they slowly made their way to Crete, stopping by several islands. These included Santorini, an island that is the site of the largest volcanic eruption in recorded history, witnessed in 1646 BCE. "The caldera is filled with water from the Aegean Sea," wrote Hoffman. "As you sail in, you are looking up 1,000 foot cliffs."

They took a sample in the middle of the caldera and in an area called Iron Bay, so named because its waters are filled with iron bubbling up from a thermal vent. "You can smell the sulfur, and the water temperature goes from 24 to 32 degrees Celsius [75 to 90 Fahrenheit]," said Hoffman. After a stop and a sample taken near Crete on September 20, *Sorcerer II* headed back to Spain, arriving on September 27 in Barcelona after seven rough days at sea. "Lots and lots of rolling around, very little sleep, high seas, and strong winds," wrote Hoffman. Since leaving Valencia on May 17, 2010, the team had traveled over seventy-two hundred miles and had collected fifty samples from six countries.

"On November 22 we took off for the USA USA USA!" wrote Hoffman as they headed from the Canary Islands to the Caribbean. "There was one problem, though: a huge storm in the Northern Atlantic. To avoid the storm, we had to go much more south than originally planned. Also this storm sucked all the wind up north, so we had very little wind to sail with. With no wind and a much longer sail than planned, we couldn't make it directly to Florida . . . . Well, we could have but we would have run out of fuel and food."

On December 11, *Sorcerer II* finally arrived in St. Thomas in the US Virgin Islands, and six days later in Florida. The total catch for Expedition II: 213 samples taken from the waters of thirteen countries, two oceans, and five major seas (plus eight mahi- mahis, three wahoos, and two yellowfin).

AFTER EUROPE, the sampling expeditions continued, some of them off *Sorcerer II*. In early 2014, a team headed by Jeff Hoffman traveled to the Amazon River basin in Brazil. This was following up on a trip that Craig originally wanted to take in 2004 as part of the circumnavigation. That effort had been scuttled when Friedman was told that Brazil was not keen on foreigners collecting microbes from their waters. By 2014, the biopolitical tensions had lessened, although JCVI was still not allowed to remove any samples including DNA from Brazil. JCVI would be allowed to use the DNA codes in digital form once they had been sequenced in Brazil.

In mid-February 2014, Hoffman flew to Brazil and met up with two Brazilian collaborators: biologist Guilherme Oliveira and bioinformatician Sara Cuadros, both from the Centro de Excelência em Bioinformática (CPqRR/Fiocruz) of Belo Horizonte, situated about 272 miles north of Rio de Janeiro. The arrangement was for a joint effort to collect samples along seven hundred miles of the great river, starting in the city of Manaus upriver and ending at the city of Macapá at the river's mouth. Along the way they would take water from all five of the Amazon basin's major rivers: the Negro, Solimões, Madeira, Tapajós, and Xingu.

On February 18, 2014, the scientists arrived in Manaus, a city of two million people, and the capital of the Brazilian state of Amazonas. Founded in 1669, it sits in the middle of Earth's largest rain forest, situated where the Negro River's black water and the Solimões River's muddy brown water converge to form the Amazon. Weirdly, the two rivers run side by side, black and brown, for almost four miles before finally mixing. The city is mostly modern-looking with tall buildings, bridges, and beaches—and slums around some of the edges. It also has a sprinkling of historic buildings, mostly from the nineteenth century, when this was a booming center of the rubber industry.

According to Hoffman, writing in an unpublished log he kept, things went sideways almost the moment they arrived in Manaus, with the trip devolving at times into a comedy (and tragedy) of errors, obstacles, actors good and bad, and just plain crappy luck that he dubbed the "Amazon Shit Show."

For starters, the five-day sampling trip ended up taking seventeen days, with delays beginning when the scientists arrived in Manaus, but their equipment did not. It appeared three days later, but only after constant calls by Guilherme Oliveira and Sara Cuadros to the company that was charged with supplying what they needed for their expedition. Finally, wrote Hoffman, "a lawyer from their Institute called somebody high up and said: 'We are a government institute, and we have a contract with you. If this gear doesn't arrive ASAP, you will be blackballed by the government and won't be used.' AMAZINGLY the gear arrived on the next night!"

With their equipment finally in hand, the scientists discovered to their dismay that the ship they had hired, called *Special K* had been sold during the three-day equipment delay. The new owner promptly kicked the crew and the scientists off the boat, although Hoffman wasn't there when this happened. When he later arrived where the boat had been docked, it was gone, along with his passport, computer, and other gear. "Guilherme and I got on a fast boat (well not so fast)," wrote Jeff, "and an hour later we found the boat fueling. The owner said we can get our stuff when they get to the dock. So, we followed them. I went into my room and all my stuff was fine, so I packed up and headed back to the small boat."

That weekend the intrepid scientists secured a new, much larger boat with a crew of ten, seven of which Hoffman said seemed to be along for the ride because they didn't really do anything. On that Sunday, February 23, they took samples in the Negro and Solimões Rivers and headed back to Manaus for the night.

The team planned to depart Manaus to travel downstream the next morning at first light, although it took the crew until 4:00 PM to set off. "Then the electrical system on the boat went bonkers," wrote Hoffman. "With light bulbs exploding." Among the casualties was Jeff's smartphone, which he had been charging using the ship's electrical system. The electric compressor for taking microbial samples also "got fried," forcing the team to find a new one with a battery the next day in a town downriver about two hundred miles from Manaus called Parintins—a compressor that, of course, didn't work.

This trip, with hints of a *Heart of Darkness* scenario in comedic reverse, continued with pretty much everything imaginable going wrong—more electrical problems, ham-handed efforts to fix things, attempts to get new boats that ran into roadblocks, and on and on.

When the team finally left Parintins five days later, they took a shortcut from the Amazon River to the Tapajós River, another Amazon tributary. "It was one of the best parts of the trip, and just really felt like we were in the rain forest," reported Hoffman, "with lots of wildlife including monkeys jumping tree to tree." But the disasters continued as the scientists worried that the electrical system would start a fire. In the river town of Santarém, they started looking for a new boat. This made the captain's partner furious, because he didn't want to lose the money promised him for the trip. He started to yell at Cuadros, prompting Hoffman—a big guy—to step up and tell the man to chill out and to take them to a dock where they were going to get off the boat.

The new boat was owned by a man named Philippi, whom Hoffman deemed a "good guy." "They instantly moved all our gear, supplies and personal stuff to one of their boats," wrote Hoffman. "The team then prepared for the rest of the trip down-river while Cuadros sent the first frozen samples back to her institute to be sequenced.

The scientists continued to take samples, even though the two men were forced to sleep on mattresses on the deck because their air-conditioned room was filled with fuel fumes. They were attacked by bugs, and the heat and humidity meant wretched nights with scant sleep. When Cuadros' room, which had AC, turned out to be filled with cockroaches, spiders, and other bugs, they found yet another boat—newer, and with AC, although there was no hot water for showers. They hung in there, though, and finished collecting samples in the river and in the mouth of the Amazon.

During their last sampling, in the miles-wide mouth where the river pours into the sea, there was still another surprise, as the new boat they were on was boarded by the Brazilian navy. Unfortunately, the captain had left his license in a town up the river. The Brazilian sailors forced the boat to dock in the Naval Station in Manaus and ordered Hoffman and the scientists off the boat with all their gear. This created a dilemma because they had no dry ice to keep the samples cool, and the company that was supposed to have arranged this didn't show up. When Hoffman contacted them, they claimed to know nothing about the ice. Cuadros got the manager of their hotel in Macapa, a city near the mouth of the Amazon, to store the samples in the hotel's freezer until an alternate source of ice was found. The dry ice arrived two days later, having been delayed by the start of the Carnival holiday in the city.

"All of this for science," said Hoffman.

The scientists departed Macapa on February 25. Guilherme Oliviera and Sara Cuadros returned to their institute in Belo Horizonte, carrying the frozen samples with them, as arranged under their permit with the Brazilian government. A relieved

Hoffman headed back to San Diego. He had already spent time in Belo Horizonte teaching Olivera, who operated the institute's sequencing facility, the exact process that JCVI was using to extract and to shotgun sequence DNA, to make the methods consistent with other samples taken since the Sargasso Sea in 2003. After everything that went wrong in Brazil, the Brazilian scientists ended up doing a spectacular job and emailed the digital sequences to JCVI, with the two teams working together to do the analysis.

∀

JUST BEFORE and after the Amazon, JCVI expeditions exploring the microbiome of the planet continued, mostly on board *Sorcerer II* but not always. These included the two additional trips to Antarctica in 2012 and 2015.[18] Another JCVI expedition in 2016 explored hot, deep-sea vents in the Pacific, while scientists took numerous smaller trips off the west coast of North America, in the Caribbean, and up and down to the east coasts of North America.

The *Sorcerer II* project combed the planet for microbes until 2018, when the ship was finally retired from sampling expeditions. In all, from 2003 to 2018, the great vessel traveled over sixty-five thousand nautical miles questing for new life as it explored the Pacific, Indian, and Atlantic oceans, and the Baltic, Mediterranean, Aegean, and Black seas, extracting samples from the seas and waterways of some thirty-three countries and territories.

Scientists collected a total of 477 samples from onboard the ship, including 147 taken during the circumnavigation, and an additional 218 samples off the boat—a treasure trove of Earth's smallest life still only partially cracked open, sequenced, and reassembled. Millions of microbes remain frozen in huge metal freezers at JCVI in La Jolla waiting to be sequenced, to be analyzed, and to have their secrets revealed.

And the collecting and sequencing of samples was just the beginning of the process. What came next was to analyze what these microbes actually do—the mechanics of how they live and die, where they fit into eco-niches and microbial communities, and what their roles are in the planetary cycles of life and death.

We already have described the findings from the 2003 Sargasso Sea expedition that started it all. Now let's return to that blistering hot day on the Sea of Cortes in March 2007 when *Sorcerer II* was sailing in the vicinity of the tiny fishing village of Loreto, Mexico. That's when the crew got word that the first cache of papers about the global expedition was being launched into the world as the *PLoS Biology* special collection. It was an effort that a large team of scientists at JCVI and beyond had been working on since the first forty-one samples had been plucked out of the oceans from Halifax and Sargasso to the Galapagos.

So, what did they find?

PART III

OUTCOMES

7

A Peek into Near Infinity

The known is finite, the unknown infinite.
—THOMAS HENRY HUXLEY, ON READING
ON THE ORIGIN OF SPECIES

OFF THE COAST OF PANAMA IN JANUARY OF 2007, Doug Rusch was feeling queasy. A scrappy computational scientist with a thin face and short brown hair, Rusch was first author of the main study about the global expedition about to be published by *PLoS Biology*. Craig had invited him to join the crew of *Sorcerer II* for a few days to see what it was like to be on the open sea collecting the samples he had been helping to analyze on a computer at JCVI for most of the last two years.

Standing beside Rusch on this cloudy night on the way to San Diego was Aaron Halpern, another computational biologist at JCVI and a coauthor of the paper in *PLoS Biology*. Halpern was fumbling with the filters, trying to read instructions about how to conduct a sampling and also feeling a bit nauseated as the boat rolled and bobbed. Like Rusch, Halpern had not spent any time collecting ocean microbiome samples, or on boats. Both scientists had just received a crash course on how to bring in and filter seawater and were gamely giving it a go on a night when Jeff Hoffman and other experienced sampling scientists were not on board.

Further forward on *Sorcerer II,* captain Charlie Howard worked the throttle of the ship's engines to keep the ship in place as Rusch and Halpern continued to deploy the pump and instruments somewhat ham-handedly. "There we were in choppy waters with all of these diesel fumes," remembered Rusch. "We were sort of hanging half-upside down trying to maneuver the equipment. We were collecting off the port side of the boat and throwing up on the starboard. We finally finished filtering somewhere around 4:00 AM."

"I thought those guys should see and experience for themselves where the samples they were working with on their computers had come from," said Craig later, with a chuckle.

✔

RUSCH AND HALPERN were safely back home in Maryland when the "Ocean Metagenomics Collection" was released on March 13,

2007.[1] Their paper, "The *Sorcerer II* Global Ocean Sampling Expedition: Northwest Atlantic through Eastern Tropical Pacific," was part of it.[2] The issue—with a cover showing a close-up of the deck of *Sorcerer II* healed over amidst surging blue-gray waves— also bundled nine other articles, studies, commentaries, and essays (one added in 2008) that collectively assessed the first forty-one samples taken from the Sargasso Sea and Halifax to the Galapagos. In total, these samples had produced an eye-popping 7.7 million genetic sequences containing 6.3 billion base pairs, plus 6.12 million proteins and some seventeen hundred totally unique large protein families. In a unique meta-search for viruses, JCVI scientists had also identified 154,662 viral peptide sequences and viral scaffolds.

The special issue started with several introductory pieces describing the how and the why of the expedition, and a bit about the challenges—scientific, logistical, and political. These included a feature written by science writer Henry Nicholls called "*Sorcerer II:* The Search for Microbial Diversity Roils the Waters," describing Craig's vision for the voyage and the controversies and politics surrounding the expedition over issues like biodiversity and permits. Nicholls mentioned, for example, the trouble in countries like Bermuda and Ecuador in the Galapagos, where critics were suspicious of Craig's motives in taking samples of microbes inside their borders. "This kind of research strays into unknown biological, legal, and ethical territory," he wrote. "And in this environment, allegations of biopiracy are almost inevitable. This, however, is unlikely to deter a man like Venter. 'If it's in the Darwin school of biopiracy, then fine,' he says."

Another introductory piece, "Environmental Shotgun Sequencing: Its Potential and Challenges for Studying the Hidden

World of Microbes," was written by the well-known microbiologist Jonathan Eisen.[3] A former scientist at JCVI who is now a professor of evolution and ecology at the University of California Davis, Eisen wrote a piece describing and endorsing shotgun sequencing as the technology that made a global ocean survey possible. Eisen compared it to the invention of the Internet—"a global portal for looking at what occurs in a previously hidden world." Also like the Internet, he wrote, "there is certainly some hype associated with ESS [Environmental Shotgun Sequencing] that gives relatively trivial findings more attention than they deserve." For Eisen, however, this was outweighed by the "revolutionary potential" of the technology.

After these short pieces came the article by Doug Rusch, Aaron Halpern, and thirty-six coauthors, including Craig as senior author—a grand overview of what the expedition found in those first forty-one samples. This wasn't easy, in part because the sheer volume of data collected was unprecedented. "We started with basically a ton of information," said Halpern. "But what did it all mean? Nobody really knew. When we started, we weren't even really sure what questions we wanted to ask."

"It was a beast to put together," said Rusch, in part because the long strings of genetic code produced by the shotgun sequencing and reassembly process hadn't been compiled to test a certain hypothesis or answer specific questions. "What we had was all of this code that was out there in nature that we needed to try and make sense of, to try to ask and answer questions about the ecology and evolution from all of these different places."

Shibu Yooseph, a JCVI data scientist and molecular biologist who also wrestled with this data, remembered the unexpected challenge it created for the scientists. It was a problem they had also faced with the Sargasso data, but now even more so: "the astounding diversity of species contained in the samples. Really, it was diversity, diversity, diversity. It's the big thing that always popped up with this data."

All this diversity made it tough to assemble full genes and genomes—to piece together all the small, random fragments that resulted from "blowing up" genomes and then relying on known sequences from DNA libraries and other algorithmic tools to reassemble them. This task was made harder by the fact that microbes rearrange genes through horizontal gene transfer and not through reproduction. "In nature, microbes are all competing for resources in different and changing environments," said Yooseph, "meaning they are mutating constantly and making subtle changes in their code to better adapt—making it harder still to know how to assemble them."

Some clues for the reassembly of the shotgunned DNA came from where Craig and collaborating scientists chose to take samples. "Different populations of microbes were associated with different environmental parameters," said Rusch, "whether that be temperature or salinity, or the distance from the coast. Levels of nutrients in the water were also a big factor."

"One important finding," added Halpern, "was that, while many samples contained the same or similar organisms, they differed in subtle ways, with subtypes within subtypes of species indicating functional differences even below the subtype level." This suggested an almost infinite number of tiny evolutionary fluctuations. "We were able to use metagenomics to find hints of a sub-diversity, developing a bunch of ways of looking at an almost completely new type of data, much of which was not trivial and not obvious. But we just scratched the surface, and the discussion section of the paper is full of conjectures rather than definitive findings."

The scientists saw so much diversity in the microbial worlds they investigated that the standard taxonomy of kingdom, phylum, and species didn't capture the richness. "Talking about subtypes was an attempt to articulate that there was diversity below the species level," said Halpern. "It was structured, but the rules governing that structure were hard to determine."

"Subtypes show that there is stable diversity that is below the level of lumping and splitting that is normally done," said Rusch, "but we didn't know how diverse most of these subtypes were, or if each individual cell was literally a Frankenstein mosaic of diverse pieces, or something in between."

"We would discuss whether species were a 'cloud' of organisms, but that probably wasn't the best analogy," added Halpern. "Like, the tree of life has big branches you see from far away, and those branch into a bunch of twigs. But on the ends of the twigs are these big, fuzzy tufts, like on a willow tree in the spring, each tuft fuzzy with lots and lots of hairs." The many "hairs" in Halpern's image stood for all the diversity within subtypes.

"In that analogy, a subtype is a tuft," he continued. "We showed that even within the tuft there was a huge amount of diversity in terms of differences between the sequences of any given gene. On the other hand, we found that, despite that diversity, subtypes were relatively stable in terms of their gene content. Maybe 80 to 90 percent of the gene content was shared across subtypes. And for the genes they shared, the amounts they differed by at the sequence level were correlated."

"This meant that, for the most part, subtypes were evolutionarily distinct," said Halpern, "and they were not furiously trading bits and pieces back and forth. But for the ten to twelve percent where genes were not shared by all subtypes, these tended to be hypervariable, and showed signs of horizontal gene transfer," resulting in what Rusch described as branches of the evolutionary tree that have "an infinite breadth of small variations."

Rusch and Halpern talked about the photosynthetic bacterial species *Prochlorococcus marinus* as an example. "We had a lot of it, and in any given bucket, we had an almost infinite micro diversity," said Halpern. "We were able to show that this was true for every gene in the *Prochlorococcus* genome. Before, there had been hints of this kind of micro diversity, but I think we really established it." Rusch compared different species of *Prochlorococcus* to the difference between different primates—orangutans, chimpanzees, humans—except that you'd have to imagine primates having an almost endless variety of small differences among subtypes. "We practically never saw exactly the same thing twice," said Rusch.

"For us, this raised questions like *why* is there so much diversity?" recalled Halpern. "Is it because each of them has their moment in the sun? Their little evolutionary niche that protects them from the others? Or is it just kind of a big soup, with so many individual bacteria from any given species or subtype, with seemingly infinite diversity that can evolve and coexist?"

"Evolutionary theory suggests that all individuals in a population belonging to the same 'species' can be traced back to a common ancestor," he said, and "whether due to survival of the fittest or to dumb luck, known as genetic drift, any genetic diversity in the population has to have arisen since that common ancestor. The larger the population, the longer ago the most recent common ancestor lived, and the more genetic diversity is expected." In other words, more mutations arise across longer periods of time.

With that classic theory in mind, Halpern continued, "one way to explain our observations is to say that the size of the populations of the abundant marine microbes are so large that the common ancestor lived very long ago. However, there are other possibilities. One is geographic isolation: if a species is split between two locations that are physically isolated from one another, then separate populations can persist as genetically distinct lineages, leading to higher diversity between the populations than would be true if the populations were fully mixed." At this point, Craig chimed in: "This is what Darwin described with the unique species he found in the Galapagos." But, he emphasized, "another possibility is that there were, in fact, multiple 'common ancestors' with multiple unique origins."

"Notably, we saw high degrees of diversity within the same sample," said Halpern, "indicating that this diversity can't be solely due to geographic separation. Another possibility is that what looks like a single population in fact is two or more populations of similar but functionally distinct individuals. Because they are functionally distinct, they don't compete with one another, at least not as directly, and they can coexist, allowing more genetic differences to arise than would be expected in a single, homogeneous population. For instance, two bacteria might differ in the temperature at which they grow optimally, due to small differences in their proteins."

As expected, the most staggering diversity was demonstrated by discoveries in extreme environments, like the hypersaline pond on Floreana Island in the Galapagos. This was sample 33, taken in early February 2004 on a blistering hot day when Hoffman and his team hauled equipment inland up to this small body of water nestled into a volcanic crater. Rusch noted that, between the Floreana sample and, say, the sample taken in the freshwater Gatun Lake in Panama, there was a profound difference in salinity, supporting quite distinct sets of species with different "hairs" on the tuft-like profusions of subtypes.

Another extreme sample, 32, was taken in the coastal mangrove on February 11, 2004, in the Galapagos, when Craig jumped out of the Zodiac and sank into muck up to his knees. "The thick organic sediment at a depth of less than a meter is the likely cause for the genes and organisms we found being unlike any other sample we studied," said Rusch, "with diversity almost off the charts."

FOR THE NEXT *PLoS Biology* paper, Craig wanted to go beyond DNA to take a crack at the proteins in the first forty-one samples. The result was "The *Sorcerer II* Global Ocean Sampling Expedition: Expanding the Universe of Protein Families."[4] Computational biologist Shibu Yooseph was first author and Craig was senior author. Coauthors included Rusch and Halpern, and Shannon Williamson, Jonathan Eisen, and Karla Heidelberg, among others. At the time, and even now, this was a groundbreaking study in terms of the volume of proteins identified in the ocean and the geographic breadth of the sampling, stretching halfway around the world.

"Doug Rusch led the effort to talk about the sampling from the *genetics* side from the North Atlantic through the Galapagos," said Yooseph. "We used the same data set in our study to look at things from a *protein* perspective"—proteins being molecules that are made according to instructions from a gene and perform critical functions in an organism. "So we had all this sequence data, which allowed us to figure out the protein-coding genes and then ask some fundamental questions about how many protein families were there, and the new families that were popping up."

"We found something on the order of twenty-five hundred new protein families," said Yooseph, families that showed extensive variation from all the different sites based on location, salinity, and so forth. The researchers also found 3,995 clusters of proteins that might be families, out of which 1,700 had no detectable homology linking them to families in existing databases. This refuted the belief by many scientists that all protein families had already been discovered.

As Yooseph and the team confronted millions of unknown proteins, one of the major challenges was to try to identify how they fit into what was known and not known previously about ocean proteins and what they did. They were able to identify only about 25 percent of the total protein sequences they were working with, and the rest remained mysteries. "Which I guess was pretty good," he said, "given all that we didn't know when we started."

One curious finding was that bacteria on the ocean surface tended to have smaller genomes than those living deeper in the sea. "These guys on the surface have been really successful at adapting to an open ocean environment," said Yooseph. "They just go about doing their business. They don't have a lot of baggage with them. These bacteria also are photosynthetic and produce oxygen; they are at the bottom of the food chain that supports the ecosystem that supports us. So, if you knock some of them out, you could have all kinds of unintended consequences."

This contrasts with the organisms that are in low abundance, he said: "things like your *Vibrios*"—that is, the various bacteria of the genus that includes the comma-shaped pathogen cholera. "These tend to have a lot more extra genes that can help them survive when the conditions are not optimal. This difference in the number of genes potentially has implications for the food chain and climate change and everything else, because while they have adapted well to living in the oceans, they don't have a lot of metabolic flexibility. They do what they do really well, but if the conditions change, if the oceans become acidic suddenly, this can potentially have a negative impact on these microbes."

Y

BACTERIA OFTEN GRAB most of the attention among the different microbes that inhabit the invisible realm of Earth's oceans. At the other end of the attention spectrum are a class of microbes that are mostly ignored or vilified. They also happen to be the most abundant genetic elements on Earth. These are *viruses,* tiny bundles of DNA or RNA that may or may not be alive. These were the subject of another major paper in the *PLoS Biology* special collection, reporting on a study that was, at the time, almost unique in being a large-scale investigation into the viruses that exist in the oceans. These number in the quadrillions (10^{15}), in the greater context of an estimated ten nonillion viruses (10^{31}) existing on Earth.[5]

"Most people think 'ewww, viruses, they're bad,'" said former JCVI virologist and marine biologist Shannon Williamson, the first author of "The *Sorcerer II* Global Ocean Sampling Expedition: Metagenomic Characterization of Viruses within Aquatic Microbial Samples."[6] "But they are actually super instrumental in so many ways."

Perhaps most important are those viruses that are crucial for controlling populations of their host species. These hosts include bacteria, which proliferate so fast that, without special viruses called *bacteriophages,* their numbers would quickly overwhelm the equilibrium that allows the ecosystem that supports us to exist. Bacteriophages are also critical to carbon recycling and nutrient cycling in the ocean; when they infect and kill bacteria, they start the process of decay that breaks down and releases chemicals from the bodies of their bacterial host cells. These become nutrients to feed living bacteria, algae, and other organisms. For us humans and almost every other organism, viruses have been an important element in evolution as the DNA of ancient viruses has merged over the eons with our cellular DNA. DNA that originally came from viruses makes up about eight percent of *Homo sapiens* DNA.[7]

"When our *PLoS* paper came out, it hadn't been that long since we learned that viruses were important components of the marine ecosystem," said Williamson. "That was one of the things that drew me to study viruses in the first place as a graduate student. But the *Sorcerer* project was certainly my first experience with using genomics and metagenomics to study populations of viruses from all over the world."

"Metagenomics has been used here and there in other studies in smaller communities or in smaller oceanic communities," she said, "but just not to the extent that we were able to do for the *Sorcerer II* project. One of the interesting discoveries, but one that was also frustrating, was that we found upwards of 99 percent complete genetic novelty. Meaning that, when we would compare our data from the water to what was known—those viruses that had been sequenced and put in the genetic database, which is how a lot of the analysis was done—we looked for similarities and we found practically none." Almost everything that Williamson and her team discovered was new to science.

Viruses are exceptionally effective killing machines in the oceans, dispatching up to forty percent of the entire microbial biomass in the seas every day.[8] Viruses also attack and kill the algae raging in out-of-control blooms. No organisms are spared from attacks—viruses infect every living thing in every known ecosystem.

About nine thousand virus species have been characterized since Russian botanist Dimitri Ivanovsky first discovered a virus that was infecting tobacco plants in 1892—and the number is rapidly rising. Technically, viruses only become viruses when they enter a cell. Before that, they are *virions*, organic particles that have been described as being almost alive, but not quite, since they lack certain cellular structures and the capacity for independent replication that biologists associate with life.

Viruses are much smaller than bacteria, with diameters between twenty and three hundred nanometers. They also vary in shape. (Recently, the shape of one virus in particular became notorious for its spherical shape and surface spikes: a bug called SARS-CoV-2.) Once a virion penetrates a cell, it forces the cell machinery to rapidly produce thousands of copies of itself, a viciously simple and devastatingly efficient process. Many viruses cause cells to burst open and release new copies to infect other cells.

Scientists don't know when viruses first appeared on Earth, but some theorize they might have evolved from bacteria, or from plasmids, which are chunks of DNA that can move between cells—or might have evolved from the complex molecules and nucleic acids that existed around the time that cells first developed, some four billion years ago. The word comes from *virulentus,* Latin for poisonous. The adjective *viral* was first used in 1948 (with "going viral" on TikTok or Twitter arriving sometime later). Bacteria hardly sit idle in response to this ferocious bactericidal assault. Over billions of years these ever-resourceful organisms have developed sophisticated "immune-like" systems that aggressively attempt to fend off their tiny assailants. Part of the bacterial defense network led to the discovery of CRISPR, the gene-editing tool based on bacteria's ability to use special DNA-cutting enzymes to edit out potentially damaging viruses. "You have this constant little battle going on," said Williamson, "where the host makes modifications to avoid infection and viruses try find a way around it. Because they mutate quite quickly, and it's all trial and error. Some things make them more effective, and some things make them less effective."

Williamson described how the team on *Sorcerer II* collected and prepared the virus samples. It started with the viruses mostly being captured by the smallest filter size, 0.1 microns. "But assembly for viral data was much more difficult than with bacteria," she said, "because it was all so diverse. So even though viruses make up the most abundant component of seawater—there are ten viruses for every bacterial cell in the ocean—they're so genetically diverse that we never had enough of any one particular viral species to make a complete genome back out of it. This made reference-mapping to known genomes in the database really challenging. We'd assemble what we could to get context for some of the functions, and to be able to perform phylogeny a little bit easier—to determine who is related to who, and how distantly related they are."

"It's come a long way since then," she continued, talking about progress in the field since 2008, "with more and more sequencing of virio-plankton from the oceans. There's also more data out there to compare to."

The newness of what Williamson and her team were doing circa 2008 prompted scientists at JCVI to create a new tool they called single-virus genomics. "This tool allowed us to pluck out individual viruses from an environmental sample," said Williamson, "so we could sequence its whole genome through amplification and then increase the library of reference genomes. It was the *Sorcerer* study that inspired us to pursue that development, because the analysis was so challenging and frustrating at times."

The Williamson team's first and most obvious finding was the sheer number of viruses in the samples. "It became really clear early on that viruses were truly the most diverse and most abundant biological entities on the planet," she said, confirming "something hinted at in other studies."[9] The vast majority of the viruses were bacteriophages. "We saw far fewer viruses that might infect larger macroorganisms like fish and humans."

An unexpected discovery came after the genetic analysis of the viruses suggested the samples might have been accidentally contaminated by bacteria or other organisms—in which case, some of the sequences tagged as viral genomes could actually include genes from bacteria unrelated to the viruses. "But this wasn't the case," said Williamson. "In fact, this DNA was part of the viruses that they had captured from the bacteria." It was DNA that the bacteria used to produce energy and to perform other critical functions, which the viruses had taken on.

"At the time the traditional thinking about viruses in the environment, and the marine environment in particular, was that they just carried essential genes for infection replication," Williamson continued, "and that they didn't carry around a lot of extra DNA or other genetic material. But we started seeing genes that we would normally see in larger organisms, that were involved in different metabolisms like nitrogen metabolism and photosynthesis. We realized that the viruses were invading bacteria and delivering bacteria-like genes into their host. They used these genes to keep their host on life support for a longer period of time by boosting its energy absorption and its replication efficiency." In other words, this insidious hijacking of genes helped the viruses keep their victims alive longer, giving them more time to replicate.

"Turns out viruses are awesome at shuttling genes around," she added. "They can infect a host, pick up some of its genes—say, for nitrogen metabolism—and then they replicate, burst out of the cell, go infect another host, and they have the possibility of dropping those genes off. So, they're kind of like little UPS drivers for that nitrogen gene that could, in the end, make those bacteria more successful." Clearly, if a virus killed off its host cell, it could no longer exist. This is a constant balancing act of keeping hosts alive so viruses can replicate and control the bacterial populations.

For viruses, it turns out timing is everything—timing, that is, of the exact moment they encounter a host and what the conditions are at that magic instant. "It's not a minute later, or two minutes," said Williamson, "but the bacteria the viruses are attacking change constantly, so which viruses appear where is based on the life cycle of the host. Factors like how long does it take a particular host to double its population, its growth rate." Very subtle shifts in temperature, salinity, and the rest also drive an intricate, vast, and potentially deadly dance between viruses and hosts which can change by the second, making it crazy difficult for scientists to track or understand broad patterns as these tiny clusters of DNA and RNA infect cells, mutate, and reinfect other cells.

The *Sorcerer* effort was criticized by some for taking random samples all over the world rather than taking repeated samples at the same site over a period of time. The latter approach makes pattern identification easier, allowing for more predictability of which bacteria and which viruses show up when. "But these random samples did give us a sense of just how much is out there that we didn't know about," said Williamson. "It was a starting point." She credited the *Sorcerer* effort with inspiring people to do larger studies using metagenomics, which have become easier to manage as the technology improved and sequencing got cheaper. It also helps that there are "so many more metagenomic datasets out there for viruses" today, that "comparisons to other studies are now very helpful."

⋎

EVEN AS *PLoS Biology* released its 2007 Ocean Metagenomics collection, Craig and his team at JCVI were in the midst of another colossal upload of the raw genomic data used for the analyses into public databases. This included the digital transfer of the 7.7 million genes that the JCVI teams had discovered in the bacterial samples, plus the thousands of viral peptide sequences and viral scaffolds, and the nearly six million proteins characterized by Shibu Yooseph's group.[10]

But what Craig and company placed into the public domain was not just all this molecular data. They also uploaded metagenomic data (including temperature and salinity measurements the location where each sample was collected, and more) into a brand-new database and analytical tool called the Cyberinfrastructure for Advanced Marine Microbial Research and Analysis (CAMERA).[11] This unique database was developed by JCVI in collaboration with the new computer and IT center at UCSD called Calit², which had received a grant for the purpose from the Moore Foundation.

CAMERA was the subject of the final paper in the *PLoS Biology* special collection, first-authored by JCVI computational biologist Rekha Seshadri.[12] There it is described as "a state-of-the-art computational resource with software tools to decipher the genetic code of communities of microbial life in the world's oceans." Seshadri and her coauthors outlined how this new resource would "help scientists understand how microbes function in their natural ecosystems, enable studies on the effect humans are having on the environment, as well as permit insight into the evolution of life on Earth."

"Metagenomics has the potential to shed light on the genetic controls of these processes by investigating the key players, their roles, and community compositions that may change as a function of time, climate, nutrients, carbon dioxide, and anthropogenic factors," the article continued. "One can envision a future where metadata from satellites and weather stations, and other physicochemical data, can be used to help interpret and inform scientists on how these factors affect microbial processes as well as community composition."

CAMERA's technology was built using a new computing technology developed at Calit[2] and several other universities and institutes, called the "OptiPuter." At the time, this project was prototyping a global-scale version of an end-to-end cyberinfrastructure that could generate a "high-resolution visualization cluster" of images and data originating in a researcher's lab and transfer it over dedicated one- or ten-gigabit-per-second optical fibers to remote data and computer servers anywhere. In the early 2000s, this was a new and radical concept.

The leader of the CAMERA project was the founder and first director of the UCSD center, Larry Smarr.[2] An astrophysicist who had helped design the OptiPuter technology, Smarr years earlier had shifted from physics to computers and helped to develop a network of supercomputers in the 1980s that was an early predecessor to the Internet.

Smarr's involvement with CAMERA began in 2005 when he got a phone call out of the blue. "I was walking across the UC San Diego campus when Craig called me," he later remembered. "I didn't really know him, but there he was, telling me they needed to have some place to store and provide details about all the data coming out of the *Sorcerer* expedition. He thought Calit^2 was the place to do this. He was right."

Larry Smarr is a tall, bear-like man who has a child's excitability about projects and ideas he's passionate about. He has no formal training in microbiology. But he did make a pivot into the field thanks to Craig—and because of an insight he had when he teed up astrophysics next to microbiology. These are two disciplines that seem at first glance to be about as far apart as one can imagine—pitting the unfathomable bigness of studying the universe against the unfathomable smallness of peering into the microbiome. But Smarr said that, as he got into the CAMERA project and learned what Craig and others were doing, he made the following observation: "there are maybe 10^{12} galaxies and a hundred billion or so stars, so 10^{22} stars in the universe. And then I learned that on Earth, this little flyspeck in the middle of nowhere, out in the boonies of the universe, there are 10^{30} bacteria. So, a *hundred million times* as many bacteria as there are stars in the universe. And it's just mind-boggling, right? And so I realized that the ultimate data that's going to change everything in science is going to be microbe biology. And that's why I've been working in it since then."

Not long after Craig's call, Larry Smarr literally put his gut into the effort by starting what became a years-long project to measure and keep track of changes in his own microbiome. He did this by collecting a stool sample every other day and having JCVI and others sequence the microbes. He tracked the data and changes using the massive computing power he has access to at Calit^2. At one point, Smarr's self-testing revealed a disturbing change in his microbiome—the early onset of Crohn's disease, which he fended off by carefully adjusting his diet and lifestyle, all the while deploying the full force of the cutting-edge technology at his disposal to track his progress. In a way, this was a space guy turning his telescope inward, not only to study his own constellation of trillions of bacteria, but to change it to his advantage.

"We ended up with over six thousand users of the infrastructure, from seventy-something countries," said Smarr about the CAMERA project. "It became this common watering hole that was available to everybody interested." Unfortunately, in 2014, CAMERA's funding ran out and it was shut down. "In part, this happened because the project was too successful," said the Moore Foundation's David Kingsbury. "There was more demand than we had anticipated," he said, "and we couldn't really keep up and meet it the way people wanted us to. We just didn't have the resources to do it. Whenever you make grants, you always worry about what are you going to do if it fails. What we rarely do is plan for *what are we going to do if it really works?*" Fortunately, as Smarr noted, others have created databases and tools like CAMERA: "It's now just what everybody does."

⊻

WHEN THE *PLoS Biology* collection came out in 2007, Craig and Kenneth Nealson wrote a summary piece about it in *ISME Journal* in which they emphasized how much was still missing. The Global Ocean Survey, they noted, focused on the "near-surface marine planktonic niche" and hardly captured the enormity of microbial life throughout the rest of the sea, not to mention the soil and air.[13] Neither did it include very small animals and larger phytoplankton (eukaryotes) and other larger microbes plus multicellular bacterial mats, attached cells, and symbionts. "In many cases," they wrote, "these bacteria have well-defined niches ... and there is no doubt that they play a role in marine ecosystems."

"Each of us should sit down with the data and add our private interests and expertise to its analyses," they concluded. "The important thing now is to seize the moment and move forward."

This is exactly what has happened since 2007 as microbiologists and oceanographers all over the word have tapped into the millions of microbes scooped up and sequenced by *JCVI* scientists, way beyond those first forty-one samples taken from the Sargasso to the Galapagos. Chapter 8 will get into the crush of data and analysis since the *Sorcerer II* project began in 2003, which has generated thousands of papers and experiments that have certainly pushed forward the science of environmental microbiology. It will also offer some thoughts about how this explosion of discovery has helped us better understand where humans fit into a planet dominated by microbes.

This includes a gathering realization that, historically and biologically, we are not as in control of our planet as we have thought, and that humans are just one more organism deeply connected to and dependent on a planet of microbes that don't care about our species, except as one of many niches of resources to inhabit and work with—sometimes symbiotically and sometimes not. This newfound humility comes even as we also learn about the significant impact of human activity on the environment at a global scale. Both realizations compel us to stop pretending the microbiome doesn't matter.

8

More Microbes than Stars

It goes on forever—and—oh my God!—it's full of stars!
—ARTHUR C. CLARK, *2001: A SPACE ODYSSEY*

ON A LATE FALL AFTERNOON IN 2018, near La Jolla, California, JCVI microbiologist Chris Dupont stood on the beach by the Scripps Institution of Oceanography, preparing to enter the cold waters of the Pacific Ocean. Stepping into the waves crashing onto the sand, he carried a long surfboard and a satchel containing a precleaned, two-liter sample bottle. Hopping on his board, the thirtyish Dupont, a compact man with an easy smile, paddled out to a prescribed location offshore to collect a sample of ocean water filled with microbes. It's something he's been doing off this beach for years, combining, like Craig, two of his passions: surfing and the study of ocean microbes.

Dupont was conducting one of the tens of thousands of focused and detail-oriented experiments that researchers have been doing since the *Sorcerer II* circumnavigation, producing discoveries and insights that have helped to illuminate the science of what we are learning about the ocean microbiome. "This was a small study," said Dupont, like dozens he has engaged in since joining JCVI in 2009, and earlier in his career. "But hopefully it helped in the wider effort in a smallish way."

DuPont was gathering seawater for a project he was working on with Paul Carini, a microbiologist at the University of Arizona. The two scientists were collecting samples from the ocean here in southern California to add to soil samples taken near Tucson in Arizona's Santa Catalina Mountains. The aim was to try to better understand why some microbes in these two environments—soil and ocean—are easier to culture in a laboratory than others, an effort that included shotgun sequencing to check the DNA of the microbes they were studying.

"I got myself into a good forward glide," recounted Dupont, "opened the bottle, rinsed it with seawater, then filled it. I capped it and put it back into the satchel." He then paddled to where the waves were cresting and caught one particularly sweet one back to the beach, where he put the sample into a cooler and drove back to his lab. "I filtered half for sequencing at JCVI and put the rest to send to my collaborator's lab in Arizona," said Dupont.

"They were doing the same with the soil samples they were collecting, sending half to me and keeping half." In early 2020, the team published a study on the soil part of the project.[1] "We're still working on publishing the ocean half," said Dupont.

Dupont's work, like that of many other microbiologists in the past decade and a half, has been focused on teasing out insights from samples taken on *Sorcerer II* and by scientists working on other microbe-gathering projects large and small—including a few on surfboards. A small selection of Dupont's output, often in collaboration with Craig and other JCVI scientists, includes a paper published in 2011 on fluctuations of phytoplankton in the sea off the coast of San Diego.[2] Another article, published in 2015, researches how genomes and gene expression differed in a variety of levels of sunlight and nutrient availability at different depths in the eastern, subtropical Pacific.[3] In 2017, Dupont published an analysis on the diversity of viruses discovered in the Baltic Sea, a study coauthored by Karolina Ininbergs, the Swedish microbiologist who accompanied *Sorcerer II's* 2009–2010 Expedition II in the Baltic and the Mediterranean seas.[4]

Individual papers like these have proliferated into hundreds per year. Right now, researchers are adding to the analysis of what shapes the diversity of microbes across different environments. They are discovering how specific microbes interact with each other and with other species, including humans. They are identifying the functions of different genes and proteins. And they are tracking the impacts of environmental changes on large global systems, like the marine microbial ecosystem that feeds and supports those phytoplankton that produce more than forty percent of the oxygen in the atmosphere.

Many of these studies would not be possible without another major scientific development since 2003—the insights championed by Craig that using metagenomic methods could capture a large majority of microbes that had eluded identification and classification before, and that these microbes could be collected and sequenced on a grand, global scale. This way of thinking helped to inspire other multiyear global sampling expeditions.

Thinking big also encouraged large-scale public efforts to systematically study the Earth's microbiome. One of these was the Earth Microbiome Project Consortium, launched in 2010 and led by UC San Diego microbiologists Rob Knight and Jack Gilbert.[5] In 2017, that consortium published a massive study in *Nature* that sought to standardize criteria for classifying microbiome taxa and address other pressing challenges in how to characterize microbes, and to create some order in the burgeoning field of environmental microbiology.[6] The study engaged more than five hundred scientists, and 27,751 samples were acquired from forty-three countries. "These samples represent myriad specimen types and span a wide range of biotic and abiotic factors, geographic locations, and physicochemical properties," wrote Knight and Gilbert in a 2018 editorial published in *mSystems* about the study.[7]

The study, said Knight and Gilbert, offered scientists a chance "to test fundamental hypotheses in biogeography, including determining patterns that have previously been possible only for 'macrobial' ecology. In addition, the ecological trends demonstrated key organizing principles whereby ecosystems with less diversity maintained taxa that were found in samples with greater diversity." The data also allowed researchers to "explore the factors that underpin global diversity trends" and, by using informatics techniques, they were able to "highlight the local adaptation and therefore environmental specificity of subspecies." This represented an extension of the analysis of the genetic and proteomic diversity of the first forty-one samples taken from *Sorcerer II,* as reported by JCVI papers in the 2007 *PLoS Biology* special issue.

What the Earth Microbiome Project didn't do, added Knight and Gilbert, was to explore the function of genes or shed light on other molecular processes, like which proteins are expressed by these genes and the role of metabolites in these organisms. The study of such phenomena is known as *multi-omics* because it combines several "omics" like genomics, proteomics, micro-biomics, and metabolomics. In recent years, multi-omics has become all the rage in molecular biology, although it remains a fantastically complex task to factor in so many dynamic processes all at once.

"To assess how microbes are distributed across environments globally—or whether microbial community dynamics follow fundamental ecological 'laws' at a planetary scale—requires either a massive monolithic cross-environment survey or a practical methodology for coordinating many independent surveys," wrote Knight and Gilbert. They also made the critical point that, even as samples and studies pile up, the question of what it all means remains a work in progress. Scientists are simply overwhelmed by data—a curious bookend to the scarcity of data that prevailed pre-2003.

⩒

ONE WAY TO think about microbial ecosystems on Earth—and their impact on humans—is to imagine something akin to a Russian matryoshka doll. That's a doll within a doll within a doll, with the outermost layer in this case being the ecosystem doll we see from space, the one that's spherical and colored blue, green, and white. This huge ecosystem, which supports life as we know it, contains all the other ecosystems within ecosystems within ecosystems, all the way down to the micro-ecosystems in the sea, deep in the Earth, or on the pistil of a rose—or inside of you.

The human microbial ecosystem is merely one out of an almost infinite number of sub-matryoshkas on Earth. Yet because the human microbiome is rather important to most of the organisms reading this book, we'll use it as an example. In a book about the microbiome of the sea and the Earth, we'll take a moment to describe how the *Sorcerer II* project and environmental microbiology over the last twenty years has contributed to our understanding of the trillions of tiny organisms—eukaryotes, archaea, bacteria, viruses, and fungi—that live inside and on *Homo sapiens*.

The latest estimate puts the average population of microbial life inside us at around thirty-nine trillion.[8] The number of bacterial cells roughly equals the number of much larger human cells.[9] Microbes collectively carry five hundred to a thousand times the number of genes we have in our human cells, but they account for less than one pound of the average person's overall weight.[10]

Our microbes mediate the basics of our organismal maintenance by performing their usual tasks of secreting chemicals, breaking down cells that die, reprocessing basic chemicals like carbon, nitrogen, and oxygen, and much more. Hundreds of millions of microbes in the gut help us digest and process food. They protect us, when in a proper balance, from disease and perform a host of functions that keep us healthy. Microbes from our mothers, acquired in part from foods eaten and from other outside ecosystems, affect our development even before birth. They break down our food to extract nutrients we need to survive.[11] They teach our immune systems how to recognize dangerous invaders and helpfully produce anti-inflammatory compounds that fight off disease-causing microbes.

An ever-growing number of studies have demonstrated that changes in the composition of our microbiomes correlate with numerous disease states, raising the possibility that manipulation of these communities could be used to treat disease—a prospect that several pharmaceutical and biotech companies are working on to develop new drugs and other treatments.

Part of this research is investigating what happens when we indulge in high-sugar and high-fat diets, sleep badly, or drink too many gin and tonics, all of which can cause "bad" bacteria to proliferate in our guts at the expense of "good" bacteria, making us sick and even affecting our moods. For instance, a 2018 study in the *European Journal of Nutrition* investigated the delicate balance between carbohydrate consumption and the level of bacteria in the gut that break carbohydrates down into metabolites. The researchers discovered that when a person cuts back too much on carbs, the bacteria that consumes them turns to other sources of food in the gut. This causes the carb-loving bacteria to excrete less healthy metabolites.[12] Another study in *Nature Microbiology* in 2019 suggested that the prevalence of certain gut microbes correlates with depression.[13]

Until recently, however, the identification and study of large systems of microbes, good and bad, in people was limited by the pesky stubbornness of most bacteria to be cultured in the lab. This is where the ecosystem of the human microbiome enters our story, and the story of how the techniques developed to sequence microbes in the environment in the late 1990s and early 2000s were used later to better understand the microbiome of humans.

"I think the study of environmental microbiomes absolutely laid the foundation for work on humans," said microbiologist Karen Nelson, former president of JCVI and an early advocate of using metagenomics to study the human microbiome. "The first series of investigative steps that used metagenomics were definitely with environmental samples, with water and with soil, in part because it was easier to go in and see what was there. We then used those same techniques on humans."

The origins of Nelson's work with human microorganisms go back to Craig's final couple of years at Celera, from 2000 to 2002. "You have to understand that Craig at Celera and TIGR had possibly the highest-throughput operation for sequencing on the planet," said Nelson. "And while everyone was focused on sequencing human DNA, we and others started to ask questions about other animals like chickens and cows—and also what was going on in their GI tracts in terms of microbes—wondering what we would find if we tried shotgun sequencing." At the time, most of the genetic work in microbiology, human or otherwise, involved 16S rRNA. "My lab was busy doing some basic 16S work with oral bacteria samples in people, with microbiologist David Relman at Stanford," said Nelson. "And before that, we had been funded to look at bacteria, including gingivitis, in the oral cavity."

According to Nelson, the idea of using shotgun sequencing on the human gut first cropped up with a project that Craig and

Ham Smith thought they might pursue at Celera. "They wanted to sequence the microbes in human poop," said Nelson, and were calling the project "HuPoo." Celera's parent company, however, decided its for-profit business wasn't the place to do what sounded like basic research. So Craig moved the project to TIGR. "I went to Craig one day and asked, 'can I take this over?' I wanted to do the whole genome shotgun sequencing that he was doing in the oceans on the gut microbiome. And he said yes, so that's what we did."

In 2006, Nelson led a team that published the first major paper in *Science* that used metagenomics to analyze the entirety of microbes found in the feces of humans, a great source of raw material (so to speak) of what's going on in the microbiome of the gut.[14] "For that paper, we were rapidly able to leverage for humans what was going on in the environmental world," said Nelson. They started by trying out their idea on the feces of just two people—an N=2 that would never fly today, since thousands of people would need to be tested for a finding to be interesting to scientists and physicians. But at the time, the whole idea of shotgunning gut bacteria was new.

"We were really surprised by the diversity of the flora in the two individuals' feces," said Nelson, "the community was stunned. No one expected that, just like no one had expected the diversity Craig found in the Sargasso Sea. We weren't quite sure what to do with it, there was so much. Because you must understand that everybody was culturing one organism or two organisms at a time back then. And now we had thousands, most of them organisms you'd never seen before."

"Then, over time, we started to look at different people," she added, with different profiles: "healthy versus disease, antibiotic

usage, and all the rest. We could also see what was called a 'dysbiosis,' an imbalance in the microbiome that can cause disease.
At first, we just didn't understand what was going on. What
caused this? Was it diet? Something else? To this day it's still
hard to see cause and effect, but you clearly see correlations
in many cases between an imbalance of certain bacteria and
disease."

Nelson gives a current-day example of microbiologists digging into feces and cracking open microbial genomes to sleuth
out the impact of a specific dysbiosis in some human guts—in
this case, one that appears to be associated with nonalcoholic
fatty liver disease. As the most common form of chronic liver
disease in the United States, this condition affects between
eighty million and one hundred million Americans, and kills
twenty thousand a year. Up to seven thousand patients get liver
transplants as a result of it. In 2017, UC San Diego gastroenter-
ologist Rohit Loomba led a study, which Craig coauthored
along with Karen Nelson, involving a clinical study of eighty-six
patients with liver disease.[15] They took stool samples and found
some likely indicators for advanced liver disease, most notably
E. coli and another bacterium called *Bacteroides vulgatus.*

Both are abundantly available in all human guts, and both
can be helpful or harmful depending on how many or how few
of them are present, and on other factors such as diet. *Bacteroides vulgatus* is part of a family of bacteria called *bacteriodes*
that are among the most populous portion of a person's microbiota, with up to 10^{11} little *Bacteroides* in our guts. They play a
crucial role in breaking down complex molecules in the food
we eat, and are especially prolific when a person consumes lots
of protein and animal fats and goes easy on carbohydrates. The

study found that overabundances of both these bacteria were associated with advanced liver disease and the authors offered a theory of how the overabundance might have spurred production of metabolites toxic to the liver. In the summer of 2021, Loomba was planning to expand his study to five hundred diabetics, diabetes being a disease that puts patients at high risk for nonalcoholic fatty liver disease.

"We're working on developing a panel of bacteria that we could use to test for and diagnose liver disease," said Loomba. Such an advance would avoid the need for a diagnostic biopsy. "It's an idea that could be used to diagnose other diseases where specific bacteria behave in a certain way that contributes to a disease. We can also use these tests to see if a treatment has worked, by checking how it affected the microflora."

Small but significant findings like this about the human microbiome over the past fifteen years have proliferated almost as wildly as, and possibly more than, the universe of environmental microbiome studies. The NIH launched and funded a big science project, the $170 million Human Microbiome Project, which was active from 2007 to 2016, with JCVI as a collaborator.[16] This initiative had echoes of the Human Genome Project, although with considerably less funding.

Primarily, the Human Microbiome Project focused on identifying bugs in humans and understanding how they affect health and disease. It involved big efforts in the sort of infrastructure-building and standardization that also was happening in environmental microbiology with the Earth Microbiome Project. When the Human Microbiome Project ended, it listed among its accomplishments the identification of some ten thousand microbial species that inhabit the human ecosystem.[17] All of

these were included in a catalog of reference metagenomes sequenced from different sites in the body—mostly using 16S rRNA,
but in some cases using whole genome sequencing, too. Related
work was also done on legal and ethical issues around privacy,
what patients should be told when, and commercialization.[18]

In specific projects, scientists studied such questions as the
microbiome of the vagina before giving birth, the viral load in
the microbiome of children with unexplained fevers, and the
role of the microbiome in diseases of the skin, digestive system,
and reproductive organs.[19] One study established a link between certain microbes and atherosclerosis.[20] There have been
many more. In 2012, *Nature*, *PLoS Biology*, and other journals
published several studies coming out of the Human Microbiome Project.[21] Scientists were treated to several surprises, including the observation that microbes contribute genes tasked
with helping humans survive. Humans' microflorae were also
seen to change over time in reaction to disease and environmental shifts, which stands to reason, given that the human
microbiome is merely one of countless ecosystems nested inside the big and intricately connected matryoshka ecosystem
of Planet Microbe. Most gut microbiomes studied, however,
tended to return to a state of homeostatic equilibrium after
fending off disease or experiencing an imbalance in microflorae,
sometimes by establishing a different balance of microbes than
before the disruption.

Given the recent all-out fight against a global pandemic, it
should also be mentioned that research into SARS-CoV-2 has
advanced the science of the human microbiome for coronavi-
ruses and other pathogenic viruses—and to some extent, for
viruses in general. Scientists from around the world joined in

an all-hands effort to understand the novel coronavirus (that's now not so novel) and how it interacts with its human hosts. As a result, we now know more about the role of human genetics in why some people experience severe symptoms and die, while others suffer no symptoms at all.[22] And we are better equipped to create new and better vaccines to shut down not only SARS-CoV-2 but the truly novel viruses that undoubtedly will appear in the future.

9

A Microbial "Inconvenient Truth"

Only within the moment of time represented by the present century has one species—man—acquired significant power to alter the nature of the world.

—RACHEL CARSON, *SILENT SPRING*

MICROBIOLOGIST CHRIS DUPONT REMEMBERS the first time he saw a sea lion go zombie berserk near La Jolla, California. "His eyes were rolling, and he was viciously attacking anything within range," said Dupont. He blamed the animal's behavior on a potent neurotoxin called domoic acid that's produced by a microbe called *Pseudo-nitzschia,* an algae that can go from benign to toxic. "It's very poisonous to sea lions that eat it," he said, "and to humans."

A likely contributor to *Pseudo-nitzschia* going domoic is the warming of the ocean off California as greenhouse gases bump up the temperature of our planet.[1] Near La Jolla, the water has increased 0.6 degrees per decade since the 1970s.[2] That rise has correlated in recent years with surges in seasonal domoic acid concentrations in the ocean that are "comparable to some of the highest values that have been recorded in the literature," according to the California Office of Environmental Health Hazard Assessment.[3] "Zooplankton and small fish eat the *Pseudo-nitzschia,*" said Dupont, "and it moves up the food chain to sea lions."

JCVI oceanographer Andrew Allen has spent years studying *Pseudo-nitzschia,* which belongs to a genus of algae known as diatoms, one of the largest phytoplankton in the oceans. He explains that warming isn't the only factor at work in *Pseudo-nitzschia* going toxic. The process is still not entirely understood, but the conversion seems to happen when these microbes gorge on high levels of nutrients in cooler water off the California coast and then, as he puts it, "they slam into warmer water near the coast"—water that has been increasing in temperature as part of climate change.

If this sounds complicated, it is. Huge global systems of currents, winds, salinity, nutrients, and temperature are at work and they combine in a constantly changing choreography of interactions with impact on, among other things, individual microbes like *Pseudo-nitzschia.* Climate change adds to the dance and to what ultimately happens to these long, cylindrical, microscopic algae—and sea lions and countless other organisms, as eight billion people have their effects on the vast and invisible multitude of bacteria, viruses, fungi, and algae in the sea. These are the 10^{29} microorganisms estimated to live in the oceans, helping to make the delicate planetary ecosystem one that sustains us humans instead of one that does not.

People are much more aware of what climate change and pollution is doing to macro life.[4] They hear about the steep declines in everything from polar bears to monarch butterflies.[5] They learn about damage to nature, from rare orchids in the Amazon basin to tricolored blackbirds in North America.[6] Yet the effect on invisible microbes like *Pseudo-nitzschia* can also be profound. To be sure, we must hastily add a caveat: unlike butterflies, birds, and humans, which vanish forever if they go extinct, short-lived microbes like *Pseudo-nitzschia* constantly mutate and evolve. In response to environmental fluctuations they are far more likely to react, adapt, and change than to completely disappear.

"Human activity is causing a huge imbalance in the global microbiome," said Craig. "Go fifteen miles South of here after it rains, and all the human waste from Tijuana runs down the Tijuana River into the ocean. It's affecting and polluting all the beaches not only in Mexico but also in Southern California with *E. coli* and other infections in the water, and people are getting sick. We're also seeing a rise in ocean acidification of about 0.1 pH since preindustrial times.[7] So yes, we're affecting the environment."

"With the melting of all the ice, the water level is going to be higher and the oceans are warming," he added. "With ocean warming of only one degree, you're getting reactions like what is happening to coral reefs, where a key bacterium that is responsible for keeping reefs healthy is dying in some places, and along with it, the coral itself."

∀

IN 1962, marine biologist and author Rachel Carson delivered a similar message about the threats of human activity to macro life in her bestseller *Silent Spring*. This eloquent ode to the beauty and the fragility of natural life was one of the first widely read warnings about the perils of humans polluting the Earth. Carson's book is broadly credited with popularizing the environmental movement that went mainstream in the 1960s and led to, among other things, the creation of the US Environmental Protection Agency in 1970 and the nation's ban on DDT in 1972.

"For each of us, as for the robin in Michigan or the salmon in the Miramichi, this is a problem of ecology," wrote Carson, "of interrelationships, of interdependence. We poison the caddis flies in a stream and the salmon runs dwindle and die. We poison the gnats in a lake and the poison travels from link to link of the food chain and soon the birds of the lake margins become its victims. . . . These are matters of record, observable, part of the visible world around us. They reflect the web of life— or death—that scientists know as ecology."[8]

"As crude a weapon as the cave man's club, the chemical barrage has been hurled against the fabric of life," she wrote, "a fabric on the one hand delicate and destructible, on the other miraculously tough and resilient, and capable of striking back in unexpected ways. These extraordinary capacities of life have been ignored by the practitioners of chemical control who have brought to their task no 'high-minded orientation,' no humility before the vast forces with which they tamper."[9]

Carson could have been talking about the human impact on the microworld, too—how it exists as an intricate fabric of life that is both fragile and resilient in the face of an onslaught of some of the same pollutants, how human-induced stressors are changing which microbes live where, and how they interact with their environment in ways that might not be healthy for a planetary ecosystem that supports humans and the macro-organisms that support us.

Given this slow-motion but accelerating carnage of macro-life, it's easy to read Carson's description about what happens on land and in rivers, to birds and fish, when "poison travels from link to link of the food chain," and to extrapolate it to what happens when human activity causes shifts in microbial populations. The effects may be equally unhealthy for robins in Michigan and salmon in the Miramichi—and our children.

Another *Silent Spring* analogy for us to consider in the 2020s is that, while some scientists in Carson's day were keenly aware of the dangers posed by out-of-control chemical use, most people were not—not until *Silent Spring* and other clarion calls got their attention. Likewise, today's human-generated threats to the ocean microbiome remain little known outside microbiology and oceanography circles. "It's hard to get people's attention on this," said Scripps microbiologist Jack Gilbert. "It's not like people are thinking very much about the microbiome of Earth."

Gilbert mentioned the 2021 Netflix film *Don't Look Up* as an portrayal of how human society can be so self-absorbed that it fails to act even when the danger is right there in sky—an asteroid poised to slam into the Earth.[10] The movie is a satire of human folly, starring Leonardo DiCaprio and Jennifer Lawrence as scientists trying to tell a world so obsessed with social media memes, likes, gossip, and what Stephen Colbert once called "truthiness" that they fail to take seriously an extinction-level event that everyone could see coming if they, well, just *looked up*. And never mind an invisible threat—a looming catastrophe that might kill us all coming from a world most people are only vaguely aware exists.

"It's hard to get the attention of politicians and others about what's happening," said Dupont. But scientists are trying. For instance, in 2019, a group of thirty-four microbiologists published a paper in *Nature Reviews Microbiology* titled "Scientists' Warning to Humanity: Microorganisms and Climate Change." The authors wrote:

> In the Anthropocene,* in which we now live, climate change is impacting most life on Earth. Microorganisms support the existence of all higher trophic life forms. To understand how humans and other life forms on Earth (including those we are yet to discover) can withstand anthropogenic climate change, it is vital to incorporate knowledge of the microbial 'unseen majority.'
>
> We must learn not just how microorganisms affect climate change (including production and consumption of greenhouse gases) but also how they will be affected by climate change and other human activities.

* *National Geographic's* online resource library defines the term as follows: "The Anthropocene Epoch is an unofficial unit of geologic time used to describe the most recent period in Earth's history when human activity started to have a significant impact on the planet's climate and ecosystems."

This Consensus Statement documents the central role and global importance of microorganisms in climate change biology. It also puts humanity on notice that the impact of climate change will depend heavily on responses of microorganisms, which are essential for achieving an environmentally sustainable future.[11]

Another way to think about this is to add the impact of human activity on the planet's microbiome to former vice president Al Gore's notion of *An Inconvenient Truth*—the title of a 2006 documentary about the perils of global warming, directed by David Guggenheim, and of Gore's 2007 book (which could be thought of as a version of *Silent Spring* for the twenty-first century).[12] The film and book highlight the rise of carbon dioxide in the atmosphere and the "inconvenient truth" that the burning of fossil fuels to power our civilizations and lifestyles is causing climate change, which will get worse if humans fail to act to reduce CO_2 emissions.

Like Carson, Gore refers mostly to macro life and changes in weather patterns and other evidence we can see with our own eyes. The micro world isn't much discussed—although, since that film and book, scientists like Craig, Chris Dupont, Andrew Allen, and Jack Gilbert (among many others) have been hard at work to reveal the impact of CO_2 buildup and pollution on Earth's smallest organisms. They have been building on efforts begun decades ago that accelerated with the deployment of technologies like shotgun sequencing and metagenomics around the turn of the twenty-first century, and with the urgings of people like Craig to think about microbes on a global scale.

Other planet-scale projects following in *Sorcerer II's* wake, like the Earth Microbiome Project, have continued the effort that is now beginning to yield a broader understanding not just of how the ocean's microbiome naturally works in concert with wind, temperature, and other huge planetary systems, but also how human activity is affecting these natural processes. Yet the task to comprehend something so big and complicated remains a huge challenge and much remains unknown, a point expressed by the thirty-four scientists who authored the 2019 "warning to humanity" in their consensus statement:

> Long-term data are needed to reliably predict how microbial functions and feedback mechanisms will respond to climate change, yet only very few such datasets exist (for example, the Hawaii Ocean Time-series and the Bermuda Atlantic Time-series Study). In this context, the Global Ocean Sampling Expedition, transects of the Southern Ocean, and the Tara Oceans Consortium provide metagenome data that are a valuable baseline of marine microorganisms.[13]

"Most of us have such a human-centric view of the world," said Craig, "like the Earth was made for us, and it will keep supporting us no matter what we throw at the environment. It's not very smart of us. We can't live at pH 11, we can't live at pH 1, we can't live in a methane atmosphere, we can't live with too much CO_2. But that isn't really given much thought by most people, or by politicians. Even some biologists tend to focus on the environment that supports humans like it's set in stone, or it's the gold standard of what life is, which is kind of disastrously wrong."

"Over increasingly large areas of the United States," wrote Rachel Carson in 1962, "spring now comes unheralded by the return of the birds, and the early mornings are strangely silent where once they were filled with the beauty of bird song."[14] It's sad that we can now add to this dire if poetic forecast that human activity is also causing Planet Microbe to change in strange ways—increasing the likelihood that the tweets, buzzes, roars, purrs, caws, oinks, and barks that we love and that nurture us will be hushed forever.

$\vee$

ONE OF THE most critical planetwide systems that ultimately ensures the birds will sing and the salmon will run—or not—is something called the ocean carbon biological pump, or just the biological pump. This is the system in the oceans that's responsible for absorbing and removing twenty-five to thirty percent of the CO_2 emitted into the atmosphere and sequestering it at the bottom of the sea.[15] The pump also supports the ocean's role in producing about fifty percent of the oxygen in the atmosphere.

What makes the biological pump work is phytoplankton.[16] These are the tiny organisms that *Sorcerer II* scientists collected from around the world—microbes that inhabit a slice of ocean near the ocean's surface where they convert sunlight into energy through photosynthesis and, like plants, absorb carbon dioxide and release oxygen. "The biological pump works because phytoplankton are photosynthesizing," said Allen, and when they die, some of them sink and take with them the carbon they have on board where it gets mired in the muck and stays there for perhaps millions of years. "We're lucky we have the oceans to sponge up so much CO_2," said Dupont. "If the pump that drives this ever stopped working, we'd be in big trouble."

Equally vital to keeping the biological pump humming along are upwellings from the bottom of the ocean that deliver critical nutrients to phytoplankton. These include the nitrates, phosphate, sulfur, calcium, iron, and magnesium that come from decaying bodies of dead phytoplankton and other sources and are pushed up toward the surface where phytoplankton live, by wind, currents, and temperature changes.

One way to understand how human activity may be affecting the biological pump is to observe the oceans from space using satellites, something the crew on *Sorcerer II* did in 2009 in places like the west coast of Mexico. They used images of phytoplankton blooms snapped from the OrbView-2 satellite that was then orbiting Earth carrying an instrument called SeaWIFS—short for Sea-viewing Wide Field-of-view Sensor. SeaWIFS picked up chlorophyll in the phytoplankton and monitored how these blooms were proliferating around cities like Puerto Vallarta and Acapulco as they were fed and overfed by human waste, nitrogen and phosphorous runoff from fertilizers, and other effluents.

Back in 2009, however, fewer details were known about how exactly pollutants were affecting the phytoplankton there, or what forces were at work with temperature changes, nutrient flows, salinity, and so on. In those days the primary mission was to collect samples and deploy metagenomics and other tools to see what was there and which microbe populations lived in what environments. This was a continuation of the work done in the Sargasso Sea and on the global ocean sampling (GOS) expedition. "A lot of our real milestones and breakthroughs have been on the heels of GOS and really doing the sort of functional genomics that grew out of it," said Allen. "I wouldn't say that these are necessarily GOS projects, but I think that genomics, metagenomics, and metatranscriptomics"—the latter being the study of the microbes' gene expression within natural environments—"have given us tools that have the sensitivity necessary to start making observations that are important to understanding these processes in more detail."

One example is Allen's research drilling down on the intricacies of domoic acid. He and others have been working to understand why this neurotoxin is proliferating off the west coast on the United States. It's a question that has become urgent for the shellfish industry because domoic acid not only drives sea lions zombie berserk, it also gets into crabs and other shellfish that people eat. A high-dose exposure to domoic acid in humans can cause seizures and short-term memory loss.

"In 2015 there was a massive domoic acid bloom on the west coast," said Allen. "It caused hundreds of millions of dollars in damage to the commercial fisheries, if you combine Washington, Oregon, and California. You had Dungeness crab, other types of shellfish, and marine mammals impacted. Now the outbreaks seem to be occurring more regularly, and we need to understand it better." In this case, the impact on industry and human health has led to funding from the US National Oceanic and Atmospheric Administration (NOAA) that is helping to fast-track the work of scientists like Allen trying to better understand the cellular and oceanographic mechanisms that drive domoic acid production.

"Warming seems to be a factor," said Allen about what they have learned so far, "although it's way more complex than that. We've found that if diatoms [like *Pseudo-nitzschia*] are growing happily and blooming offshore and then suddenly slam into warm waters near shore where there is some sort of nutrient limitation—like too little silicon—then they start producing domoic acid." It was his belief that these conditions might activate a genetic response or, in some cases, cause an increase in certain species of *Pseudo-nitzschia* that are prone to producing domoic acid. "Not all *Pseudo-nitzschia* do it," he said, "but we're starting to get a handle on some of the mechanisms that do cause it and we can reliably stimulate it in the lab."[17]

Allen adds that there is a Hollywood angle to domoic acid, which does have a diabolical-sounding name that seems plucked from a horror movie. "A lot of people believe that Alfred Hitchcock's film *The Birds* was motivated by a real-life incident where crazy birds were slamming into windows after they had eaten anchovies that had been eating domoic acid," he said, referring to an actual occurrence in the town of Capitola, California, near Santa Cruz, in the 1960s. Here is an account of this zombie berserk bird attack written up in a 2016 article in the *Mercury News:*

> In August 1961, Capitola residents awoke to a scene that seemed straight out of a horror movie. Hordes of seabirds were dive-bombing their homes, crashing into cars and spewing half-digested anchovies onto lawns.
>
> Famed film director Alfred Hitchcock even used the incident as research material for his then-in-progress movie "The Birds," in which flocks of deranged birds inexplicably attack a coastal town.

Just as the unexplainable avian attacks in "The Birds" have terrified movie buffs for more than half a century, the 1961 frenzy puzzled scientists for decades. They now believe the culprit was domoic acid—the same neurotoxin that has delayed this year's Dungeness crab season in California.[18]

In another series of studies detailing the minutia of how microbes thrive or don't, the focus is on iron, a critical nutrient for phytoplankton to be able to absorb nutrients like nitrogen from the ocean.[19] (Iron is critical for humans, too.) These studies date back to JCVI's Antarctica expeditions from 2006 to 2015, and investigations into how rising levels of carbon in the atmosphere and the acidification of the oceans might be undermining a delicate mechanism in phytoplankton that controls iron uptake. Allen and other scientists have figured out that the natural supply of iron in the Southern Ocean is limited, which means that fewer phytoplankton live there. Yet researchers also have found that increasingly warm water in the Southern Ocean seems to be substituting for iron in some phytoplankton and driving recent increases in their population.

This phytoplankton population bump, said Allen, could have implications for regions far away from Antarctica, like the southern Pacific Ocean. In the past, the fact that iron-deprived phytoplankton in the south didn't consume all the available nutrients, such as nitrogen, led to an overabundance of nitrogen in the Southern Ocean. Traditionally, ocean currents have carried this excess north, which has provided food for phytoplankton there—nourishment that doesn't happen if Antarctalian phytoplankton are proliferating and consuming more of the nitrogen before it can reach the southern Pacific.

"When you start messing with the biochemistry of the Southern Ocean, it's going to have a global pattern," said Allen, although whether this is good or bad news isn't yet known. "It's a huge challenge just trying to decipher the natural oscillations of all of this," said Allen. "And then, to understand the impacts on top of this of anthropogenically forced oscillations is even more difficult."

All this planetary oscillating is also changing the amount of phytoplankton in the oceans. What exactly is happening is unclear; some studies and models suggest declines in populations while others show increases. According to NASA, the food supply for some phytoplankton is under pressure and decreasing as greenhouse gases trap sunlight and bump up atmospheric and ocean temperatures.[20] As stated on a NASA website, "Productivity is expected to drop because as the surface waters warm, the water column becomes increasingly stratified." In other words, the upwells don't flow as easily.[21]

It is possible for phytoplankton to be both increasing and decreasing: drops in the populations of larger phytoplankton, such as diatoms, can occur at the same time that smaller types, like cyanobacteria, are increasing. Such a rebalancing of big and little could have implications for planetary carbon sequestration.[22] Size matters, since the larger phytoplankton absorb about forty percent of the carbon dioxide that ends up on the ocean floor.[23]

Allen suggested that the trend might be toward more phytoplankton overall, with most of the increase coming from the little guys. "This could lead into a situation where the ocean is not responsible for as much net oxygen production," he said. In that case it would sequester less carbon dioxide, "because you've changed the size structure and community composition of the food." This isn't a helpful trend for species that need oxygen and would like to see reductions in CO_2 in the atmosphere.

Y

LET'S SAY YOU'RE FLYING from the Baja Peninsula to Boston. Sitting in a window seat on a clear day, as you soar over the northern Gulf of Mexico, you see below the brown-green coast of Louisiana and, to the south of it, an unbroken expanse of sapphire blue sea. From the air, the sparkling water is stunning. Then you approach the mouth of the Mississippi River and note how the blue changes to a faint, light olive green that extends along the coast and out into the Gulf.

What you're seeing is another major impact of human activity on the oceans—a massive bloom of algae marking one of the larger "dead zones" in the world's oceans. This one is the size of New Jersey (almost seven thousand square miles) and growing. Also called oxygen minimum zones (OMZs), these vast patches of ocean have proliferated over the past half-century, the blooms fed by nitrogen and phosphorous chemicals that are disgorged by the world's great rivers into the sea. Mostly, these chemicals come from fertilizers used on farms and lawns that wash into streams and rivers, sometimes thousands of miles upstream, and wind up in the ocean, causing algae and other microbes that thrive on nitrogen and phosphorus to explode in growth. They also consume oxygen and block underwater plants from getting sunlight. When the algae die, more oxygen in the water is consumed by bacteria, leaving very little for fish and other macro life.[24]

"The dead zone off the coast of Louisiana is the largest it's ever been," said Dupont. "You can see it from space. It stretches from Texas all the way to the east end of Louisiana. NOAA has a live feed."[25] But the Gulf is hardly alone. These dead zones have spread from the mouths of rivers all over the planet, ranging from the coast of Mexico, where the crew of *Sorcerer II* saw blooms, to the Yangtze in China, the Nile in Egypt, and freshwater bodies inland such as Lake Erie.

Dupont likened what was happening with dead zones to what Rachel Carson wrote about when chemicals got into ecosystems and affected birds and fish far away from where the pollutants originated. "When farmers don't use fertilizers correctly and use too much, they are sending chemicals to a place that's five hundred miles away," he said. "So you get these situations where farmers using fertilizer, or factories using metals in the Midwest, become a problem for Louisiana." If that weren't bad enough, bacteria in dead zones also produce nitric oxide, which bubbles up into the atmosphere and adds to global warming.

Not that dead zones contain no life. "Certain microbes are in booming health," he explained. "They're getting fed everything they need, but they just happen to be eating up all the oxygen." Some of the bacteria in dead zones are anaerobic and don't need or want oxygen. They "breathe" things like nitrate, ammonium, manganese, iron, and sulfate. "At the very worst," said Dupont, "they'll actually use CO_2 and turn it into methane," a greenhouse gas even more associated with climate change than carbon dioxide.

Another phenomenon affecting oxygen levels is a rise in "hypoxic," oxygen deficient, pockets in the open sea, which have been increasing over the past fifty years. Andrew Allen talked about an extreme hypoxic pocket that appeared from 2014 to 2016 off the coast of California called a marine heat wave: "We got this huge lens of warm water that just sat there, unmovable."

Dead zones and open ocean hypoxic zones combined now cover more than twelve million square miles of ocean and extend as much as two hundred meters below sea level, constituting an area collectively larger than North America or Africa.[26] Overall, studies show an average seawater oxygen decrease across the globe of about two percent between 1960 and 2017.[27] The levels have declined both in the open sea and in coastal waters.[28]

Another Carsonian web-of-life dislocation needs mentioning here—the one afflicting coral reefs around the world, including some spectacular ones visited by *Sorcerer II*. This story starts with the tiny algae called zooxanthellae and bacteria that live in the tissue of coral, providing their hosts with certain important nutrients, helping to remove waste, and performing certain immunological tasks that help fend off pathogens.[29] Some of these microbes are also responsible for coral's dazzling colors. The problems come when rising temperatures, pollutants, and other natural and human-produced stressors cause the coral to eject the zooxanthellae and other symbiotic microbes.[30] This plays havoc with the intricate system of micro and macro life on reefs and causes coral to bleach white, while making it more susceptible to disease and death.[31]

ANOTHER POLLUTANT in the oceans and in the environment that was just beginning to be a problem back in Rachel Carson's day is plastics. The crews on *Sorcerer II* saw plastic everywhere, especially in the great expanses of floating plastic slippers, cartons, slurpy cups, packing crates, and much more that have collected in gyres in the open sea. An estimated five trillion pieces of plastic now litter the top two hundred meters of the oceans' surface.[32] These include an estimated twelve to twenty-one million metric tons of just three major types of plastic: polyethylene, polypropylene, and polystyrene.[33] The true extent of plastic in the oceans, including the micro-pellets less than five millimeters in size that routinely break down from larger pieces, is unknown. What is known is that the micro-pellets ingested by fish, crustaceans, and other sea animals are harmful and sometimes deadly.[34] They turn up inside of people who eat fish, too, although usually in minute quantities.[35] They may be toxic to some phytoplankton.

Because plastic is mostly nonbiodegradable, bacteria and other microbes can't break them down. That's why they're out there floating in the sea, sometimes in those continent-sized, half-sunken plastic islands that Craig and the crew of *Sorcerer II* saw in the early 2000s—and that have since become much bigger.

"Plastics are a new ecosystem in the ocean," Dupont said, "that harbors bacteria, phytoplankton, and even bigger organisms that come from the land or near coasts that are not generally found in the open ocean." He was referring to a process called "rafting." Much remains unknown about the interactions between microbes and plastic in the oceans, he emphasized: "We're still trying to figure out what exactly is going on."

Y

DEAD ZONES, drops in oxygen, the occasional killer diatom, micropellets of plastic lurking in our fish filets . . . as depressing as all this sounds, the situation doesn't have to be unremittingly bleak. There was a time in recent history in the United States and other parts of the world when humans rallied to clean up parts of the environment we had trashed. By the 1960s, Los Angeles,

New York City, and other cities were becoming too smoggy for residents to breathe, and many lakes and rivers were so toxic that people wading into them could get chemical burns. Prompted by these undeniable facts—and thanks to the activism of Rachel Carson and the environmental movement—Congress passed legislation to address the crisis.

Starting in 1963 and continuing with amendments in the 1970s and beyond, the Clean Air Act succeeded in scrubbing the air of the worst pollutants. According to the US Environmental Protection Agency, the act is responsible for a fifty percent drop in emissions of major pollutants since 1990.[36] Still, a recent "State of the Air" report released by the American Lung Association estimates that about 150 million Americans are breathing polluted air.[37]

In 1972, Congress passed the Clean Water Act, which has contributed to the cleanup of many of America's bodies of water— although, to this day, more than half of American streams and rivers, about 70 percent of lakes and ponds, and around 90 percent of coastal ocean area are in violation of the act's water quality standards.[38] To say there is more to do is not to ignore the progress already made; those of us who were around in the toxic 1970s and 1980s can confirm that by the 1990s we were able to swim and fish in lakes and streams where we had been unwilling to risk even dipping a toe.

In 2006, David wrote an article for *National Geographic* called "Chemicals within Us." The piece started with him being tested for levels of chemicals inside him (mostly present in his blood) including pesticides, heavy metals, dioxins, and other toxins that he had been exposed to over the years. Indeed, nearly all humans today carry these inside them in trace parts-per-billion and parts-per-trillion amounts.[39] As part of the story, David investigated where he might have been exposed to certain chemicals that showed up in his blood, researching places he had lived from the time he was born. He made the surprise finding that a waste dump he had grown up just a mile from had probably allowed toxins to seep into the local drinking water. In the 1980s it was declared a major EPA superfund site, placing it on a national priority list of hazardous places that needed to be cleaned up.

"I grew up in northeastern Kansas, a few miles outside Kansas City," wrote David. "There I spent countless hot, muggy summer days playing in a dump near the Kansas River. Situated on a high limestone bluff above the fast brown water lined by cottonwoods and railroad tracks, the dump was a mother lode of old bottles, broken machines, steering wheels, and other items only kids can fully appreciate. We had no idea that for years companies and individuals had also discarded thousands of pounds of toxic chemicals in our dump-playground." The article continued:

> It was started as a landfill before there were any rules and regulations on how landfills were done," says Denise Jordan-Izaguirre, the regional representative for the federal Agency for Toxic Substances and Disease Registry. "There were metal tailings and heavy metals dumped in there. It was unfenced, unrestricted, so kids had access to it."
>
> Kids like me.
>
> Now capped, sealed, and closely monitored, the dump, called the Doepke-Holliday Site, also happens to be half a mile (0.8 kilometers) upriver from a county water intake that supplied drinking water for my family and forty-five thousand other households. "From what we can gather, there were contaminants going into the river," says Shelley

Brodie, the EPA Remedial Project Manager for Doepke. In
the 1960s, the county treated water drawn from the river,
but not for all contaminants. Drinking water also came
from 21 wells that tapped the aquifer near Doepke.

"Today the air is clear," David reported, "and the river free of
effluents." The site was a visible testament to the success of the
US environmental cleanup spurred by the Clean Air and Clean
Water Acts of the 1970s. The Doepke dump where he played is
cleaned up and the toxins removed or sealed underground, although it requires constant monitoring to make sure the chemicals buried there don't leak into the environment.

Yet the question remains: Can we humans act on something
that's even more complicated and massive than what was at-
tempted with the Clean Air and Clean Water Acts—an effort
that needs to dramatically alter human behavior on a planetary
scale? In recent years, detailed and increasingly alarming studies
and reports detailing why this needs to be done have been issued by the United Nations Intergovernmental Panel on Climate
Change and other groups. At climate change conferences in
places like Paris, Madrid, Marrakech, and Glasgow, most countries have pledged to reduce CO_2 emissions in an effort to keep
global warming below 1.5 degrees.[40]

Yet actually making this happen remains tough work, tangled in politics and the exigencies of economic development—
plus resistance from certain industries, climate deniers, and
more generally a civilization still dependent on fossil fuels for
everything from the gas we burn in our cars to the plastic sunglasses we wear to the beach. Six decades after Rachel Carson
published *Silent Spring* we have to ask: Are we humans taking

seriously the "fabric of life" that is, on the one hand, "delicate and destructible, on the other miraculously tough and resilient, and capable of striking back in unexpected ways?" Are humans showing any sense of "humility before the vast forces with which they tamper?"

IN LATE 2018, the final sample was taken on board *Sorcerer II*, sample 47, in the harbor of Nantucket, Massachusetts. These final two hundred liters of ocean water were collected on a cool, misty day by Jeff Hoffman, with Charlie Howard steering the great vessel and Craig, as usual, overseeing.

In all, *Sorcerer II* as a research vessel sailed sixty-five thousand miles over fifteen years, from 2003 to 2018, collecting millions of microbes and sequencing tens of millions of genes—numbers that will continue to rise as the few last filter samples, still frozen at minus-80 Celsius in a large, silver, metal freezer at JCVI in La Jolla, are slowly thawed out and sequenced.

Visitors to JCVI can still see racks of square plastic containers filled with frozen microorganisms, in a large room off the tool shop and sequencing rooms. Open the heavy metal doors of the freezers and, after waiting for a cloud of condensed water vapor to warm up and dissipate, there they are, emerging out of the mists like something out of a science fiction movie. Their contents could be from glaciers in Antarctica, the waters off New Caledonia, lakes above the Arctic Circle in Sweden, or many other places around the world—in any case, they are samples that continue to hold their secrets.

This is the point in our story where we pause to wonder if microbes are more than just part of a vast global system that's being altered by human activity, and to consider whether they might also be solutions to the "inconvenient truth" we face as humans. Maybe an unknown microbe tucked into this freezer could offer a novel solution to global warming—a new source of clean energy or clean replacement for a key industrial chemical, fertilizer, or plastic. Or perhaps there is a tiny bacterium in which scientists could discover a magical sequence with the potential to change the trajectory of history in a way no one could anticipate.

The notion that microbes might save us has been a key talking point for Craig for years as he's rallied support and funding for studying the microbial universe. It was in part what drove him in the 1990s to sequence *Haemophilus influenzae* and several other bacteria, including cholera.[41] Years earlier, that pathogen had afflicted his fellow soldiers in Vietnam and at the Naval Medical Center in Balboa Park, San Diego, along with another he would sequence, Meningitis B.[42]

In the late 1990s Craig also began a quest to use synthetic biology to engineer new, lab-created strains of microbes, developing techniques that scientists are using to enhance the ability of natural organisms to make clean fuels, new drugs, and much more.

"We may still be saved by microbes," said Craig, referring to his work in synthetic biology, "although it's turned out to be more challenging than we originally thought."

"But that's the nature of science," he said. "It always takes longer than you think it will."

Epilogue

Thinking Bigger about Small

This is truly the "age of bacteria"—as it was in the beginning, is now and ever shall be.

—STEPHEN JAY GOULD, *THE RICHNESS OF LIFE*

EARLY IN THE *SORCERER II* GLOBAL EXPEDITION, off the coast of the Carolinas, film producer David Conover stood with Craig in the stern encouraging him to wax philosophical about the expedition and what it might find. He wanted the big picture of what Craig thought might be revealed about evolution and the role played by microbes in Earth's great web of life. In the video, Craig is wearing a bright yellow foul-weather jacket and doesn't yet have his now familiar beard. The sea and sky behind him are a moody mix of grays—a storm that will later that night roil the expedition is building up. Craig's ice-blue eyes seem to glow with a mix of determination and childlike excitement.

"We haven't had time to stop and reflect and really understand and put it all in perspective," he says, turning thoughtful. "But when I think about it, my sense is that what we're doing here is helping to accelerate what we will learn about evolution on Earth, and about our own evolution—the precise events that took place."

"As we look back at the broader pool of everything," he continues, "which started four billion years ago with microbes, I think by the end of the century we'll have the ability to understand these earlier events and how they led to us and the world around us right now. This is what some of the post human genome–sequencing experiments are starting to tell us—which is how vast the biological continuum is that we are fitting into, and how we share most of the components that have developed over time. The fine-tunings, the additions, the small changes at any given moment that collectively have led to what we see today as the differences in species."

"My guess is that, if we look at the six billion base-pair letters in our genetic code, we have basically ninety-plus percent of the same gene repertoire as other mammals. So the differences between us and chimpanzees could be as few as a hundred to a thousand changes in the genetic code." Craig's prediction, made back in 2004, would turn out to be largely true, as scientists have determined the actual difference between chimp and human DNA to be around 1.2 percent.[1]

This is, of course, humbling for a species that not long ago, in Darwin's day, had a religion-inspired belief that man was the center of everything. "You go back a few hundred years before that to see people like Galileo being threatened to be burned at the stake for just saying that the Earth was not the center of the universe," said Craig. "We have come a long way since then. Although we still think we're more important than we really are—this is why we think it's okay to be filling the atmosphere with carbon and the oceans with plastic and other garbage, like we have some kind of right to do this, even though we are very much a part of a continuum and an equilibrium that we disturb only at our own peril."

Few people think of Craig Venter as a philosopher. Primarily, he is known as a man of action who leaps headfirst into projects, often without giving much thought to what the outcome will be—like heading off on a global research expedition to collect samples of microbes not knowing what he would find. But he's also a man who consistently over the past forty-plus years has sought to unravel at a basic scientific and molecular level what life is, and how bundles of DNA go from being merely strings of nucleotides to being *alive.*

How did this happen, he has asked, and when did this occur in the evolution of life on Earth? What are the gaps in our knowledge, and how can we bridge them? What technologies can we develop to help us? And what scientific orthodoxies are wrong and getting in the way of a truer understanding?

Craig has often been blunt in making these inquiries and hasn't shied away from challenging other scientists who he believes are clinging to old ways—a habit which has won him both admirers and critics. And to be sure, there are also others inclined to challenge conventional wisdom. But you can't take away the fact that Craig Venter has asked some tough questions that go to the heart of scientists'—and philosophers'—work to understand the world and the role of humans vis-à-vis each other and the planet we live on.

"I think I have had a unique perspective in biology," said Craig to Conover's film crew in 2004. "I was the first person in history to see a complete genome of a species," referring to the bacteria *Haemophilus influenzae.* "I learned microbiology from studying the genomes of these microbes. Not by traditional methods of growing them in a lab, but by being out here, literally in the wild in a living environment, where we're tapping into this tremendous richness of diversity."

Craig is also the guy who told writer Jamie Shreeve that he wanted to sequence all the genes on Earth. This was a bit of Venter hyperbole, which even Shreeve considered to be almost embarrassingly grandiose. But did he really mean it literally? Does it matter? For Craig, it was of course attention-getting, but it also was a strident attempt to ask a big question that most scientists wouldn't dare ask, which was: "what is out there on this Planet Microbe, and what does it look like? How can it be classified, how does it work, what does it do, and how can it inform us about the basics of life and evolution and where we're headed?" Even if Craig doesn't do it himself he created the tools and proved their effectiveness for others to make it happen.

More to the point in an era of global warming, "what are humans doing to the microbiome of Earth and how is the microbiome reacting—and is there anything we can learn from microbes that might help us mitigate the effects of climate change?"

As Ari Patrinos said, "Craig practiced the kind of genomics that looks into what's going on with our planet, with our existence, at the most basic level. Especially with microbial genomics, and these critters that have been around for billions of years and have perfected all their life functions to an incredible degree."

"It's been challenging," said Craig. "People aren't very accepting of change in a rapid fashion," by which he meant the scientists who fought his ideas and tried to rein him in or sideline him. "Looking at evolution and really understanding what happened is always very frightening to a lot of people," he said. "Everything has to follow the current dogma. You can't challenge it. Even without religion being as dominant as it once was, we as a species are still afraid of the unknown and having real and sometimes surprising answers to questions."

For instance, one of the surprising answers that has emerged from the *Sorcerer II* project has to do with the role played on Earth by viruses—those tiny packets of DNA that most scientists don't consider to be alive, and that few people gave much thought to before the pandemic that raged as we wrote this book.

One of the key papers that came out of the expedition was the 2008 *PLoS Biology* paper on viruses written by Shannon Williamson.[2] "It revealed this whole world of viruses on a global scale that we didn't know much about," Craig said, "and made some findings that philosophically I struggle with. Because if you view a genome as a software system, and I would say it's the software of life, then what Shannon's study showed was that viruses aren't just agents of infection that are lethal to their hosts. In fact, they don't want to kill off their hosts, because then they have nowhere to go [to reproduce]. So, she showed that viruses in the ocean take on certain genes from the organisms they are attacking in order to keep them alive, which is pretty smart."

Williamson found that viruses attacking bacteria sometimes
bring with them a set of photosynthetic genes that they have acquired from other bacteria, which merge with the DNA of the
newly infected organism. This, in a sense, updates the software
of the new host, which keeps the host from dying and helps it to
adapt and evolve.

It's extraordinary to imagine, Craig noted, "that viruses are
programming their prey, and changing the course of evolution,
almost like we humans do in the lab when we're editing and
synthesizing genomes." It raises big questions: "Does this mean
that lowly viruses, which probably aren't even alive, are really
at the top of the chain of life? Are they in charge? If so, what
does that mean for how we view the hierarchy of life? And how
does this fit into our own evolutionary and genetic history, and
the survival of our species on Earth?"

Craig's attempts to answer these questions tend to be part
of the raw stuff of discovery, of meticulously planning logistics,
experiments, and methods, of getting one's hands dirty jumping
in Mangrove muck and yanking up samples of ocean water, of
developing technologies that can observe and break down the
components of life, of collecting and analyzing data and findings.

Darwin started with the raw stuff of discovery, too, when he
traveled to the Galapagos and elsewhere. He then settled down
to a quiet life in the UK meticulously studying his fifteen hundred samples before putting things together in *On the Origin of
Species*.

Craig hasn't been particularly quiet about his efforts to study
what the *Sorcerer II* scientists discovered in the ocean. Nor has
the effort to put things together been a solitary endeavor, as thousands of researchers have joined forces around the world to try

to unravel the secrets contained in all those microbes and all that DNA. The task also is vastly larger and more complicated, with millions of species and sub-species and trillions of genes to parse through rather than the hundreds of flora and fauna specimens that Darwin brought back to England.

"We now have better technologies than Darwin did to do the work," Craig said, "although in many ways, we are doing the exact same thing that Darwin did. We are going out and observing what is out there. But with our new tools we can look finer. We can see the genetic code, and proteins, and take sophisticated measurements. We also have supercomputers."

Darwin took twenty-three years, after his return to England, to publish *Origin*. "It's likely to take us much longer," said Craig, "to come up with any sort of universal theory or update to Darwin using what we have discovered."

"We're still trying to understand the inner relationship of the twenty thousand genes in a human, and the evolutionary steps it took to get there—or the steps that led to a sea lion or my dog to being alive and active and responsive to their environment. Right now, this is far beyond our current state of biology, and of a meaningful intellectual pursuit. All we can do at this stage is to try and look at the patterns of the new information we observe, to maybe come up with some additional principles of biology. Until then, we are still very much in a descriptive world and that's probably where biology will be for the next century at least."

Scientists also still have much more of our planet to search for microbes to sequence, and much more data remains to be gathered and analyzed and pieced together before we can create a vastly larger and more comprehensive model of how the matryoshka dolls of microbial ecosystems within ecosystems on Earth interact with other ecosystems. These include those ecosystems of microbes within and on and around us as individuals—and the vaster ecosystems that humans are impacting in inconvenient and possibly disastrous ways.

So, the contribution to science of the work described here is not a new theory of life on our planet. Craig has not written the next *On the Origin of Species*—nor is anyone else soon likely to (although you never know). But the voyage of *Sorcerer II* will still have its legacy in the science done in its wake. For an important new realm of inquiry to make progress, someone has to ask the big questions. And then get down to the work of restlessly and aggressively—at times even arrogantly—seeking out answers.

NOTES

INDEX

Notes

Prologue

1. "Large Whale Species of New England," Department of Marine Resources, State of Maine, n.d., https://www.maine.gov/dmr/science-research/species/protected /whales.html.

2. "Species Information," Department of Marine Resources, State of Maine, n.d., https://www.maine.gov/dmr/science-research/species/index.html.

3. "Harvested Seaweeds of Maine," Maine Seaweed Council, n.d., https://www .seaweedcouncil.org/identifying-maine-seaweeds/.

4. "How Much Water Is in the Ocean?" National Ocean Service, Ocean Facts, National Oceanic and Atmospheric Administration, n.d., last updated February 26, 2021, https://oceanservice.noaa.gov/facts/oceanwater.html.

5. Danielle Hall, "Marine Microbes," Smithsonian Ocean, July 2019, https://ocean .si.edu/ocean-life/microbes/marine-microbes.

6. Yinon M. Bar-on, Rob Philips, and Ron Milo, "The Biomass Distribution on Earth," *Proceedings of the National Academy of Sciences* 115, no. 25 (2018): 6506–6511, doi: 10.1073 / pnas.1711842115.

7. William B. Whitman, David C. Coleman, and William J. Wiebe, "Prokaryotes: The Unseen Majority," *Proceedings of the National Academy of Sciences* 95, no. 12 (1998): 6578–6583, doi: 10.1073 / pnas.95.12.6578.

8. Ron Sender, Shai Fuchs, and Ron Milo, "Revised Estimates for the Number of Human and Bacteria Cells in the Body," *PLoS Biology* 14, no. 8 (2016): e1002533. doi: 10.1371 / journal.pbio.1002533.

9. Jenny Howard, "Dead Zones, Explained," *National Geographic,* July 31, 2019.

10.   Howard, "Dead Zones, Explained."

11.   Patrick D. Schloss and Jo Handelsman, "Status of the Microbial Census," *Microbiology Molecular Biology Reviews* 68, no. 4 (2004): 686–691, doi: 10.1128 / MMBR .68.4.686-691.2004.

12.   E. Callaway, "'Minimal' Cell Raises Stakes in Race to Harness Synthetic Life," *Nature* 531, no. 7596 (2016): 557–558, doi:10.1038 / 531557a. This work was based on earlier experiments: C. A. Hutchison, S. N. Peterson, S. R. Gill, et al., "Global Transposon Mutagenesis and a Minimal Mycoplasma Genome," *Science* 286, no. 5447 (1999): 2165–2169, doi: 10.1126 / science.286.5447.2165; and John I. Glass, N. Assad-Garcia, N. Alperovich, et al., "Essential Genes of a Minimal Bacterium," *Proceedings of the National Academy of Sciences* 103, no. 2 (2006): 425–430, doi: 10.1073 / pnas.0510013103. Also see Daniel G. Gibson, Gwynned A. Benders, Cynthia Andrews-Pfannkoch, et al., "Complete Chemical Synthesis, Assembly, and Cloning of a *Mycoplasma genitalium* Genome," *Science* 319, no. 5867 (2008): 1215–1220, doi: 10.1126 / science.1151721; and Daniel G. Gibson, John I. Glass, Carole Lartigue, et al., "Creation of a Bacterial Cell Controlled by a Chemically Synthesized Genome," *Science* 329, no. 5987 (2010): 52–56.

1. Sargasso Sea Surprise

Epigraph: Francis Bacon, "Of Cunning," in *Essays of Francis Bacon*, ed. Mary Augusta Scott (New York: Scribners, 1908), 105.

1.   "Bermuda Atlantic Time-Series Study," Bermuda Institute of Ocean Sciences, n.d., http://bats.bios.edu/.

2.   Hydrostation "S," Bermuda Institute of Ocean Sciences, n.d., www.bios.edu /research/projects/hydrostation-s/.

3.   A. Laffoley et al., "The Sargasso Sea," Government of Bermuda, based on 2011 draft, https://oceanfdn.org/sites/default/files/EBSA_submitted+(1).compressed.pdf.

4.   J. Craig Venter, Mark D. Adams, Eugene W. Myers, et al., "The Sequence of the Human Genome," *Science* 291, no. 5507 (2001): 1304–1351, https://science.sciencemag .org/content/291/5507/1304.

5.   E. Lander et al., "Initial Sequencing and Analysis of the Human Genome," *Nature* 409, no. 6822 (2001): 860–921, doi: 10.1038 / 35057062.

6.   The quoted words are as Craig spoke them to the director of the documentary: David Conover, *Cracking the Ocean Code*, Compass Light, Discovery Communications, 2005.

7.   James Shreeve, "Craig Venter's Epic Voyage to Redefine the Origin of the Species," *Wired*, August 1, 2004, https://www.wired.com/2004/08/venter/.

8.   Bijal Trivedi, "Profile of Rino Rappuoli," *Proceedings of the National Academy of Sciences USA* 103, no. 29 (2006): 10831–10833, doi: 10.1073 / pnas.0604892103.

9.   Trivedi, "Profile of Rino Rappuoli."

10.   Shreeve, "Craig Venter's Epic Voyage."

2. Planet Microbe

Epigraph: Leon M. Lederman with Dick Teresi, *The God Particle: If the Universe Is the Answer, What Is the Question?* (Boston: Houghton Mifflin, 1993).

1.   Sarah E. DeWeerdt, "The World's Toughest Bacterium," *Genome News Network,* July 5, 2002, http://www.genomenewsnetwork.org/articles/07_02/deinococcus.shtml.

2.   Yuko Kawaguchi, Mio Shibuya, Iori Kinoshita, et al., "DNA Damage and Survival Time Course of Deinococcal Cell Pellets during Three Years of Exposure to Outer Space," *Frontiers in Microbiology* 11 (August 2020), art. 2050, doi: 10.3389 / fmicb .2020.02050.

3.   "Most Radiation Resistant Microbe," *Guinness World Records,* n.d., accessed September 7, 2021, https://www.guinnessworldrecords.com/world-records/66429 -most-radiation-resistant-microbe.

4.   "Bacterial Endospore," from Jules Berman, *Taxonomic Guide to Infectious Diseases* (San Diego: Elsevier Science and Technology, 2012), cited in Science Direct, https://www.sciencedirect.com/topics/biochemistry-genetics-and-molecular -biology/bacterial-endospore.

5.   Raul J. Cano and Monica K. Borucki, "Revival and Identification of Bacterial Spores in 25- to 40-Million-Year-Old Dominican Amber," *Science* 268, no. 5213 (1995): 1060–1064, doi: 10.1126 / science.7538699; also G. W. Gould, "History of Science: Spores; Lewis B Perry Memorial Lecture 2005," *Journal of Applied Microbiology* 101, no. 3 (2006): 507–513, doi: 10.1111 / j.1365-2672.2006.02888.x.

6.   Hans-Curt Flemming and Stefan Wuertz, "Bacteria and Archaea on Earth and Their Abundance in Biofilms," *Nature Reviews Microbiology* 17, no. 4 (2019): 247–260, doi: 10.1038 / s41579-019-0158-9.

7.   N. DeLeon-Rodriguez et al., "Microbiome of the Upper Troposphere: Species Composition and Prevalence, Effects of Tropical Storms, and Atmospheric Implications," *Proceedings of the National Academy of Sciences* 110, no. 7 (2013): 2575–2580, doi:10.1073 / pnas.1212089110.

8.   US Geological Survey, "African Dust Carries Microbes across the Ocean: Are They Affecting Human Ecosystem Health?" USGS Open-File Report 03-028, January 2003, https://pubs.usgs.gov/of/2003/0028/report.pdf.

9.   A. Checinska Sielaff et al., "Characterization of the Total and Viable Bacterial and Fungal Communities Associated with the International Space Station Surfaces," *Microbiome* 7, (2019), art. 50, doi: 10.1186 / s40168-019-0666-x.

10.   "Virulence and Drug Resistance of Burkholderia Species Isolated from International Space Station Potable Water Systems," J. Craig Venter Institute blog, 2020, https://www.jcvi.org/research/virulence-and-drug-resistance-burkholderia -species-isolated-international-space-station; M. D. Lee et al., "Reference-guided Metagenomics Reveals Genome-level Evidence of Potential Microbial Transmission from the ISS Environment to an Astronaut's Microbiome," *iScience* 24, no. 2 (January 29, 2021), doi: 10.1016/j.isci.2021.102114.

11.   Carl Zimmer, "Scientists Say Canadian Bacteria Fossils May Be Earth's Oldest," *New York Times,* March 1, 2017.

12.   Viviane Richter, "What Came First, Cells or Viruses? A Biological Enigma That Goes to the Heart of the Origin of Life," *Cosmos,* October 19, 2015, https:// cosmosmagazine.com/science/biology/what-came-first-cells-or-viruses/.

13.   David Biello, "The Origin of Oxygen in Earth's Atmosphere," *Scientific American,* August 19, 2009.

14.   Liz Thompson, "Unlocking the Secrets of Earth's Early Atmosphere," Phys.Org, Science X Network, April 28, 2021, https://phys.org/news/2021-04-secrets-earth-early -atmosphere.html.

15.   "How Much Oxygen Comes from the Ocean?" National Ocean Service, Ocean Facts, National Oceanic and Atmospheric Administration, n.d., last updated February 26, 2021, https://oceanservice.noaa.gov/facts/ocean-oxygen.html. Most microbiologists estimate the oceans produce around 50 percent of the O_2 in Earth's atmosphere.

16.   "Eukaryotes: A New Timetable of Evolution," Max-Planck-Gesellshaft, June 1, 2015, https://www.mpg.de/9256248/eukaryotes-evolution.

17.   Craig Venter, *Life at the Speed of Light: From the Double Helix to the Dawn of Synthetic Life* (New York: Viking, 2013).

18.   Michael Crichton, *Jurassic Park* (New York: Knopf, 1990).

3. The Ocean's Genome Goes Meta

1.   J. McIntyre and J. Ostell, eds., *NCBI Handbook* (Bethesda, MD: National Center for Biotechnology Information, 2002–), accessed 25 September 2022, https://www .ncbi.nlm.nih.gov/books/NBK21101/?depth=2.

2.   Nicholas Wade, "Long-Held Beliefs Are Challenged by New Human Genome Analysis," *New York Times,* February 12, 2001.

3.   F. Sanger et al., "Nucleotide Sequence of Bacteriophage φX174 DNA," *Nature* 265, no. 5596 (1977): 687–695, doi: 10.1038 / 265687a0.

4.   R. Gardner et al., "The Complete Nucleotide Sequence of an Infectious Clone of Cauliflower Mosaic Virus by M13mp7 Shotgun Sequencing," *Nucleic Acids Research* 9, no. 12 (1981): 2871–2888, doi: 10.1093 / nar / 9.12.2871.

5.   "The Nobel Prize for Physiology or Medicine 1978," Nobel Prize, https://www .nobelprize.org/prizes/medicine/1978/summary/.

6.   Craig Venter, *Life at the Speed of Light: From the Double Helix to the Dawn of Synthetic Life* (New York: Viking, 2013), 52.

7.   R. D. Fleischmann et al., "Whole-Genome Random Sequencing and Assembly of *Haemophilus influenzae* Rd," *Science* 269, no. 5223 (August 1995): 496–512, doi: 10.1126 / science.7542800.

8.   Craig Venter, *A Life Decoded: My Genome, My Life* (New York: Viking, 2007), 207.

9.   H. O. Smith, J. D. Peterson, K. E. Nelson, et al., "DNA Sequence of Both Chromosomes of the Cholera Pathogen *Vibrio cholerae,*" *Nature* 406, no. 6795 (2000): 477–483, doi: 10.1038 / 35020000.

10.   Carol J. Bult et al., "Complete Genome Sequence of the Methanogenic Archaeon, *Methanococcus jannaschii,*" *Science* 273, no. 5278 (1996): 1058–1073, doi: 10.1126 / science.273.5278.1058.

11.   Venter, *Life at the Speed of Light.*

12.   Venter, *Life at the Speed of Light.*

13.   Nicholas Wade, "Deep Sea Yields a Clue to Life's Origin," *New York Times,* August 23, 1996, 23.

14.   J. C. Venter, K. Remington, J. Heidelberg, et al., "Environmental Genome Shotgun Sequencing of the Sargasso Sea," *Science* 304, no. 5667 (2004), 66–74, doi: 10.1126 / science.1093857.

15.   Venter et al., "Environmental Genome Shotgun Sequencing of the Sargasso Sea."

16.   James Shreeve, "Craig Venter's Epic Voyage to Redefine the Origin of the Species," *Wired,* August 1, 2004, https://www.wired.com/2004/08/venter/.

17.   Andrew Pollack, "Groundbreaking Gene Scientist Is Taking His Craft to the Oceans," *New York Times,* March 5, 2004, A19.

18.   Kate Ruder, "Exploring the Sargasso Sea," *Genome News Network,* March 4, 2004, http://www.genomenewsnetwork.org/articles/2004/03/04/sargasso.php.

19.   Pollack, "Groundbreaking Gene Scientist."

20.   Pollack, "Groundbreaking Gene Scientist."

21.   Venter et al., "Environmental Genome Shotgun Sequencing of the Sargasso Sea."

22.   Pollack, "Groundbreaking Gene Scientist."

23.   L. Gómez-Consarnau et al., "Microbial Rhodopsins Are Major Contributors to the Solar Energy Captured in the Sea," *Science Advances* 5, no. 8 (August 7, 2019), doi: 10.1126 / sciadv.aaw8855. The acronym IBEA refers to the Institute for Biological Energy Alternatives, one of the research entities later folded into the J. Craig Venter Institute.

24.   M. Heinrich, "Nairobi Final Act of the Conference for the Adoption of the Agreed Text of the Convention on Biological Diversity," in *Handbook of the Convention on Biological Diversity,* ed. Secretariat of the Convention on Biological Diversity (London: Earthscan, 2001), https://www.cbd.int/doc/handbook/cbd-hb-09 -en.pdf.

25.   "Rocking the Boat: J. Craig Venter's Microbial Collecting Expedition Under Fire in Latin America," ETC Group, July 21, 2004,https://www.etcgroup.org/content /rocking-boat-j-craig-venters-microbial-collecting-expedition-under-fire-latin -america.

26.   Rex Dalton, "Natural Resources: Bioprospects Less Than Golden," *Nature* 429, no. 6992 (2004): 598–600, doi: 10.1038 / 429598a, 600, box entitled "Bermuda Gets Tough over Resource Collecting."

27.   J. Ward, A. Knap, and J. M. Short, "Bermuda Welcomes Careful Prospectors," letter to the editor, *Nature* 430, no. 7001 (2004): 723, doi: 10.1038 / 430723b.

28.   Shreeve, "Craig Venter's Epic Voyage."

29.   Pollack, "Groundbreaking Gene Scientist."

4. Halifax to the Galapagos

Epigraph: James Shreeve, "Craig Venter's Epic Voyage to Redefine the Origin of the Species," *Wired,* August 1, 2004, https://www.wired.com/2004/08/venter/.

1.   Eric Linklater, *The Voyage of the Challenger* (Garden City, NY: Doubleday, 1972), 67–69.

2.   Peter K. Weyl, *Oceanography: An Introduction to the Marine Environment* (New York: John Wiley, 1970), 49.

3.   *Report on the scientific results of the voyage of H.M.S. Challenger during the years 1873–76 under the command of Captain George S. Nares . . . and the late Cap-*

tain Frank Tourle Thomson (Edinburgh: printed by Neill and Co. for H.M.S.O., 1880–1895). Murray quote at vol. 50 (1895), "A Summary of the Scientific Results," *xii*.

4. A. Chakravarti, "Obituary: Victor Almon McKusick (1921–2008)," *Nature* 455, no. 7209 (September 4, 2008), 46, doi: 10.1038 / 455046a.

5. M. Heinrich, "Nairobi Final Act of the Conference for the Adoption of the Agreed Text of the Convention on Biological Diversity," in *Handbook of the Convention on Biological Diversity,* ed. Secretariat of the Convention on Biological Diversity (London: Earthscan, 2001), https://www.cbd.int/doc/handbook/cbd-hb-09-en .pdf.

6. "Cracking the Ocean Code," documentary, dir. David Conover, Compass Light, 2005, originally aired on Discovery Channel.

7. Douglas Rusch et al., "The *Sorcerer II* Global Ocean Sampling Expedition: Northwest Atlantic through Eastern Tropical Pacific," *PLoS Biology* 5, no. 3 (2007): 398–431, doi: 10.1371 / journal.pbio.0050077.

8. Rusch et al., "The *Sorcerer II* Global Ocean Sampling Expedition."

9. Rusch et al., "The *Sorcerer II* Global Ocean Sampling Expedition."

10. "About Us," Charles Darwin Foundation, Galapagos, n.d., https://www .darwinfoundation.org/en/about.

11. Shreeve, "Craig Venter's Epic Voyage."

12. "Playing God in the Galapagos," ETC Group, March 10, 2004, https://www .etcgroup.org/content/playing-god-galapagos.

13. Shreeve, "Craig Venter's Epic Voyage."

14. "Peter and Rosemary Grant, 2005 Balzan Prize for Population Biology," International Balzan Prize Foundation, 2005, https://www.balzan.org/en/prizewinners /peter-and-rosemary-grant/.

15. Rosemary Grant and Peter Grant, *Evolutionary Dynamics of a Natural Population: Large Cactus Finch of the Galapagos* (Chicago: University of Chicago Press, 1989); Peter Grant and Rosemary Grant, *How and Why Species Multiply: The Radiation of Darwin's Finches* (Princeton: Princeton University Press, 2008); Peter Grant and Rosemary Grant, *40 Years of Evolution: Darwin's Finches on Daphne Major Island* (Princeton: Princeton University Press, 2014).

16. Jonathan Weiner, *The Beak of the Finch: A Story of Evolution in Our Time* (New York: Vintage, 1994).

17. K. Sterelny, *Dawkins vs. Gould: Survival of the Fittest,* new ed. (Cambridge: Icon, 2007).

5. French Polynesia to Fort Lauderdale

1.   James Shreeve, "Craig Venter's Epic Voyage to Redefine the Origin of the Species," *Wired,* August 1, 2004, https://www.wired.com/2004/08/venter/.

2.   J. Craig Venter Institute, "Metagenomic Analysis of Marine Microbes Isolated during the Global Ocean Sampling Expedition," National Library of Science, March 16, 2007, http://www.ncbi.nlm.nih.gov/entrez/query.fcgi?db=genomeprj& cmd=Retrieve&dopt=Overview&list_uids=13694.

3.   Douglas Rusch et al., "The *Sorcerer II* Global Ocean Sampling Expedition: Northwest Atlantic through Eastern Tropical Pacific," *PLoS Biology* 5, no. 3 (2007): 398–431, doi: 10.1371/journal.pbio.0050077.

4.   Juan Enriquez, *As the Future Catches You: How the Genome and Other Forces Are Changing Your Life, Work, Health, and Wealth* (New York: Crown Business, 2001).

5.   Craig Venter, *A Life Decoded: My Genome, My Life* (New York: Viking, 2007).

6.   Herman Melville, *Typee: A Peep at Polynesian Life* (New York: Wiley and Putnam, 1846; London: Penguin Classics, 1996).

7.   Robert Louis Stevenson, *In the South Seas* (New York: Scribner's, 1896).

8.   All Memoranda of Understanding and other agreements concerning permits in countries visited by *Sorcerer II* are available from JCVI; the one with French Polynesia can be accessed at https://www.jcvi.org/sites/default/files/assets/projects /gos/collaborative-agreements/French_Polynesion_MOU_French.pdf.

9.   "This Is Gump Station," Gump Station, University of California, n.d., https:// moorea.berkeley.edu/.

10.   Shreeve, "Craig Venter's Epic Voyage."

11.   R. Cavicchioli, M. Z. DeMaere, and T. Thomas, "Metagenomic Studies Reveal the Critical and Wide-ranging Ecological Importance of Uncultivated Archaea: The Role of Ammonia Oxidizers," *BioEssays* 29, no. 1 (2007): 11–14, doi: 10.1002 / bies.20519; Anne Mai-Prochnow et al., "Hydrogen Peroxide Linked to Lysine Oxidase Activity Facilitates Biofilm Differentiation and Dispersal in Several Gram-Negative Bacteria," *Journal of Bacteriology* 190, no. 15 (2008): 5493–5501, doi: 10.1128 / JB.00549-08.

12.   Examples: F. Ballestriero et al., "Identification of Compounds with Bioactivity against the Nematode *Caenorhabditis elegans* by a Screen Based on the Functional Genomics of the Marine Bacterium *Pseudoalteromonas tunicata* D2," *Applied and Environmental Microbiology* 76, no. 17 (2010): 5710–5717, doi: 10.1128 / AEM.00695-10; Catherine Burke, Staffen Kjelleberg, and Thomas Torsten, "Selective Extraction of Bacterial DNA from the Surfaces of Macroalgae," *Applied and Environmental Micro-*

biology 75, no. 1 (2009): 252–256, doi: 10.1128 / AEM.01630-08; Catherine Burke et al., "Bacterial Community Assembly Based on Functional Genes Rather Than Species," *Proceedings of the National Academy of Sciences of the United States of America* 108, no. 34 (2011): 14288–14293, doi: 10.1073 / pnas.1101591108.

6. Questing Distant Seas (Further Explorations)

Epigraph: John Steinbeck, *The Log from the Sea of Cortez* (New York: Viking, 1951; London: Penguin Modern Classics, 2001).

1.   "Oceanic Metagenomics Collection," PLOS Collections, Biology and Life Sciences, updated May 21, 2019, https://collections.plos.org/collection/ocean-metagenomics/.

2.   "Oceans Are the Real Lungs of the Planet, Says Researcher," UNESCO News, September 10, 2019, last updated April 21, 2022, https://en.unesco.org/news/oceans -are-real-lung-planet-says-researcher.

3.   P. G. Falkowski, "The Role of Phytoplankton Photosynthesis in Global Biogeochemical Cycles," *Photosynth Research* 39, no. 3 (March 1994): 235–258, doi: 10.1007 / BF00014586.

4.   Bruce V. Bigelow, "In Latest Expedition, J. Craig Venter Partners with Life Technologies," *Xconomy,* March 19, 2009, https://xconomy.com/san-diego/2009/03 /19/in-latest-expedition-j-craig-venter-partners-with-life-technologies/.

5.   Angelica Peebles, "Illumina Aims to Push Genetics beyond the Lab with $200 Genome," *Bloomberg,* September 29, 2022, https://www.bloomberg.com/news /articles/2022-09-29/illumina-delivers-200-genome-with-new-dna-sequencing -machine.

6.   Jeff Hoffman, "Sampling in Helgoland: A Warm German Welcome for the *Sorcerer II,*" J. Craig Venter Institute, June 7, 2009, https://www.jcvi.org/blog/sampling -helgoland-%E2%80%94-warm-german-welcome-sorcerer-ii.

7.   Erik Mellgren, "Venter Institute Gets $8.8 Million in Stimulus Funding," *Xconomy,* June 23, 2009, https://xconomy.com/san-diego/2009/06/23/venter-institute-gets-88 -million-in-stimulus-funding/.

8.   Peter Rejcek, "Stormy Weather," *Antarctic Sun,* United States Antarctic Program, National Science Foundation, October 3, 2014, https://antarcticsun.usap.gov /features/4076/.

9.   To access JCVI blogs on the Antarctic expeditions, start with Jeff McQuaid, "Why Antarctica, and Why Now?" J. Craig Venter Institute blog, October 28, 2009, https://www.jcvi.org/blog/why-antarctica-and-why-now.

10.   Jeff McQuaid, "High Impact Science in Antarctica," J. Craig Venter Institute blog, February 28, 2009, https://www.jcvi.org/blog/high-impact-science-antar ctica.

11.   Charmaine Ng et al., "The Metaproteogenomic Analysis of a Dominant Green Sulfur Bacterium from Ace Lake, Antarctica," *ISME Journal* 4, no. 8 (2010): 1002–1019, doi: 10.1038 / ismej.2010.28.

12.   D. R. Smith et al., "Massive Difference in Synonymous Substitution Rates among Mitochondrial, Plastid, and Nuclear Genes of Phaeocystis Algae," *Molecular Phylogenetic Evolution* 71 (2014): 36–40, doi: 10.1016 / j.ympev.2013.10.018.

13.   M. Wu et al., "Manganese and Iron Deficiency in Southern Ocean *Phaeocystis antarctica* Populations Revealed through Taxon-specific Protein Indicators," *Nature Communications* 10(2019), art. 3582, doi: 10.1038 / s41467-019-11426-z.

14.   D. E. Duncan, "How Artificial Life Could Change Real Businesses," *Fortune,* May 27, 2010.

15.   "President Obama Asks Bioethics Commission to Study Implications of Synthetic Biology," Presidential Commission for the Study of Bioethical Issues press announcement, May 20, 2010, https://bioethicsarchive.georgetown.edu/pcsbi/node /81.html.

16.   D. G. Gibson et al., "Creation of a Bacterial Cell Controlled by a Chemically Synthesized Genome," *Science* 329, no. 5987 (2010): 52–56, doi: 10.1126 / science .1190719.

17.   Ian Sample, "Craig Venter Creates Synthetic Life Form," *Guardian,* May 20, 2010.

18.   "Science on the Sea Ice Edge," J. Craig Venter Institute blog, March 30, 2015, https://www.jcvi.org/blog/science-sea-ice-edge.

7. A Peek into Near Infinity

Epigraph: "Professor Huxley on the Reception of the 'Origin of Species,'" in Charles Darwin, *The Life and Letters of Charles Darwin,* ed. Francis Darwin (New York: Appleton, 1887), 1: 557.

1.   "Oceanic Metagenomics Collection," PLOS Collections, Biology and Life Sciences, May 21, 2019, https://collections.plos.org/collection/ocean-metagenomics/.

2.   Douglas Rusch et al., "The *Sorcerer II* Global Ocean Sampling Expedition: Northwest Atlantic through Eastern Tropical Pacific," *PLoS Biology* 5, no. 3 (2007): 398–431, doi: 10.1371 / journal.pbio.0050077.

3.   J. A. Eisen, "Environmental Shotgun Sequencing: Its Potential and Challenges for Studying the Hidden World of Microbes," *PLoS Biology* 5, no. 3 (2007): 384–388, doi: 10.1371 / journal.pbio.0050082.

4.   S. Yooseph et al., "The *Sorcerer II* Global Ocean Sampling Expedition: Expanding the Universe of Protein Families," *PLoS Biology* 5, no. 3 (2007): 432–466, doi: 10.1371 / journal.pbio.0050016.

5.   Katherine J. Wu, "There Are More Viruses Than Stars in the Universe. Why Do Only Some Infect Us? *National Geographic,* April 15, 2020.

6.   Shannon J. Williamson et al., "The *Sorcerer II* Global Ocean Sampling Expedition: Metagenomic Characterization of Viruses within Aquatic Microbial Samples," *PLoS One* 3, no. 1 (2008), e1456, doi: 10.1371 / journal.pone.0001456.

7.   Carrie Arnold, "The Non-human Living Inside of You," *Nautilus,* January 9, 2020, Cold Spring Harbor Laboratory website, https://www.cshl.edu/the-non-human-living -inside-of-you.

8.   "Bacteriaphage in Nature," UC San Diego Health Newsroom, n.d., https:// health.ucsd.edu/news/topics/phage-therapy/pages/phage-101.aspx.

9.   Williamson et al., "The *Sorcerer II* Global Ocean Sampling Expedition."

10.   Williamson et al., "The *Sorcerer II* Global Ocean Sampling Expedition."

11.   "Launching the Global Community Cyberinfrastructure for Advanced Marine Microbial Research and Analysis (CAMERA)," J. Craig Venter Institute, Collaborator Release, March 13, 2007, https://www.jcvi.org/media-center/launching-global-com munity-cyberinfrastructure-advanced-marine-microbial-research-and.

12.   R. Seshadri et al., "CAMERA: A Community Resource for Metagenomics," *PLoS Biology* 5, no. 3 (2007): 394–397, doi: 10.1371 / journal.pbio.0050075.

13.   K. Nealson and C. Venter, "Metagenomics and Global Ocean Survey: What's In It for Us, and Why Should We Care?" *ISME Journal* 1, no. 3 (2007): 185–187, doi: 10.1038 / ismej.2007.43.

8. More Microbes than Stars

Epigraph: Arthur C. Clark, *2001: A Space Odyssey* (London: Hutchinson, 1968).

1.   R. Bartelme et al., "Influence of Substrate Concentration on the Culturability of Heterotrophic Soil Microbes Isolated by High-Throughput Dilution-to-Extinction Cultivation," *mSphere* 5, no. 1 (2020), doi: 10.1128 / mSphere.00024-20.

2.   A. J. Lucas et al., "The Green Ribbon: Multi-scale Physical Control of Phyto-plankton Productivity and Community Structure over a Narrow Continental

Shelf," *Limnology and Oceanography* 56, no. 2 (2011): 611–626, doi: 10.4319 / lo.2011 .56.2.0611.

3.   C. Dupont et al., "Genomes and Gene Expression across Light and Productivity Gradients in Eastern Subtropical Pacific Microbial Communities," *ISME Journal* 9, no. 5 (2015): 1076–1092, 10.1038 / ismej.2014.198.

4.   L. Z. Allen et al., "The Baltic Sea Virome: Diversity and Transcriptional Activity of DNA and RNA Viruses," *mSystems* 2, no. 1 (2017), e00125, doi: 10.1128 / mSystems .00125-16.

5.   Earth Microbiome Project, http://www.earthmicrobiome.org.

6.   L. Thompson et al., "A Communal Catalogue Reveals Earth's Multiscale Microbial Diversity," *Nature* 551, no. 7681 (2017), 457–463, doi: 10.1038 / nature24621.

7.   J. Gilbert et al., "Earth Microbiome Project and Global Systems Biology," *mSystems* 3, no. 3 (2018): e00217, doi: 10.1128 / mSystems.00217-17.

8.   Mun-Keat Looi, "The Human Microbiome: Everything You Need to Know about the 39 Trillion Microbes That Call Our Bodies Home," *BBC Science Focus Magazine,* July 14, 2020, https://www.sciencefocus.com/the-human-body/human-microbiome/.

9.   Ron Sender, Shai Fuchs, and Ron Milo, "Revised Estimates for the Number of Human and Bacteria Cells in the Body," *PLoS Biology* 14, no. 8 (2016): e1002533, doi: 10.1371 / journal.pbio.1002533.

10.   Sender et al., "Revised Estimates."

11.   "Human Microbiome Research Has Massive Potential for Health Applications," J. Craig Venter Institute blog, June 27, 2019, https://www.jcvi.org/blog/human -microbiome-research-has-massive-potential-health-applications.

12.   I. Rowland et al., "Gut Microbiota Functions: Metabolism of Nutrients and Other Food Components," *European Journal of Nutrition* 57, no. 1 (2018): 1–24, doi: 10.1007 / s00394-017-1445-8.

13.   P. Strandwitz et al., "GABA-modulating Bacteria of the Human Gut Microbiota," *Nature Microbiology* 4, no. 3 (2019): 396–403, doi: 10.1038 / s41564-018-0307-3.

14.   S. R. Gill et al., "Metagenomic Analysis of the Human Distal Gut Microbiome," *Science* 313, no. 5778 (2006): 1355–1359, doi: 10.1126 / science.1124234.

15.   R. Loomba, "Gut Microbiome-Based Metagenomic Signature for Non-invasive Detection of Advanced Fibrosis in Human Nonalcoholic Fatty Liver Disease," *Cell Metabolism* 25, no. 5 (2017): 1054–1062, doi: 10.1016/j.cmet.2017.04.001.

16.   P. Turnbaugh, R. Ley, M. Hamady et al., The Human Microbiome Project," *Nature* 449, no. 7164 (2007): 804–810, doi: 10.1038 / nature06244.

17.   "NIH Human Microbiome Project Defines Normal Bacterial Makeup of the Body," National Institutes of Health news release, June 13, 2012, https://www.nih.gov

/news-events/news-releases/nih-human-microbiome-project-defines-normal
-bacterial-makeup-body.

18.   "The Human Microbiome Project," PLOS Collections, Biology and Life Sciences, May 24, 2018, https://collections.plos.org/collection/hmp/.

19.   "NIH Human Microbiome Project Defines Normal Bacterial Makeup,"

20.   O. Koren et al., "Human Oral, Gut, and Plaque Microbiota in Patients with Atherosclerosis," *Proceedings of the National Academy of Sciences of the United States of America* 108, supp. 1 (2011): 4592–4598, doi: 10.1073 / pnas.1011383107.

21.   Barbara A. Methé, Karen E. Nelson, Mihai Pop, et al. (Human Microbiome Project Consortium), "A Framework for Human Microbiome Research," *Nature* 486, no. 7402 (2012): 215–221, doi: 10.1038 / nature11209; Curtis Huttenhower et al., "Structure, Function and Diversity of the Healthy Human Microbiome," *Nature* 486, no. 7402 (2012), 207–214, doi: 10.1038 / nature11234; Karoline Faust, J. Fah Sathirapongsasuti, Jacques Izard, et al., "Microbial Co-occurrence Relationships in the Human Microbiome," *PLOS Computational Biology* 8, no. 7 (2012): e1002606, doi: 10.1371 / journal.pcbi.1002606; Sahar Abubucker et al., "Metabolic Reconstruction for Metagenomic Data and Its Application to the Human Microbiome," *PLoS Computational Biology* 8, no. 6 (2012)e1002538, doi: 10.1371 / journal.pcbi.1002358; Nicola Segata et al., "Composition of the Adult Digestive Tract Bacterial Microbiome Based on Seven Mouth Surfaces, Tonsils, Throat and Stool Samples," *Genome Biology* 13, no. 6 (2012), R42, doi: 10.1186 / gb-2012-13-6-r42.

22.   D. E. Duncan, "What Do Your Genetics Have to Do with Your Chances of Dying from Coronavirus?" *Vanity Fair,* April 16, 2020.

9. A Microbial "Inconvenient Truth"

Epigraph: Rachel Carson, *Silent Spring* (Boston: Houghton-Mifflin, 1962), 5.

1.   Z. Zhu et al., "Understanding the Blob Bloom: Warming Increases Toxicity and Abundance of the Harmful Bloom Diatom Pseudo-nitzschia in California Coastal Waters," *Harmful Algae* 67 (2017): 36–43, doi: 10.1016 / j.hal.2017.06.004.

2.   "Coastal Ocean Temperature," California Office of Environmental Health Hazard Assessment, February 11, 2019, https://oehha.ca.gov/epic/impacts-physical -systems/coastal-ocean-temperature.

3.   Jayme Smith et al., "A Decade and a Half of Pseudo-nitzschia spp. and Domoic Acid along the Coast of Southern California," *Harmful Algae* 79 (2018): 87–104, doi: 10.1016 / j.hal.2018.07.007.

4. United Nations, "UN Report: Nature's Dangerous Decline 'Unprecedented'; Species Extinction Rates 'Accelerating,'" UN Sustainable Development Goals, May 6, 2019, https://www.un.org/sustainabledevelopment/blog/2019/05/nature-decline-unprecedented-report/.

5. "Animals Affected by Climate Change," World Wildlife Fund, Fall 2015, https://www.worldwildlife.org/magazine/issues/fall-2015/articles/animals-affected-by-climate-change.

6. Kristen Pope, "Plant and Animal Species at Risk of Extinction," Yale Climate Connections, March 24, 2020, https://yaleclimateconnections.org/2020/03/plant-and-animal-species-at-risk-of-extinction/.

7. Ken Caldeira and Michael E. Wickett, "Anthropogenic Carbon and Ocean pH," *Nature* 425, no. 6956 (2003), 365, doi:10.1038 / 425365a.

8. Carson, *Silent Spring,* 189.

9. Carson, *Silent Spring,* 297.

10. *Don't Look Up,* dir. Adam McKay, Netflix, 2021.

11. R. Cavicchioli et al., "Scientists' Warning to Humanity: Microorganisms and Climate Change," *Nature Reviews Microbiology* 17, no. 9 (2019): 569–586, doi: 10.1038 / s41579-019-0222-5.

12. *An Inconvenient Truth,* documentary, dir. David Guggenheim, 2006; Al Gore, *An Inconvenient Truth: The Crisis of Global Warming,* rev. ed. (New York: Viking, 2007).

13. Cavicchioli et al., "Scientists' Warning to Humanity."

14. Carson, *Silent Spring,* 103.

15. Cavicchioli et al., "Scientists' Warning to Humanity."

16. Rebecca Lindsey and Michon Scott, "What Are Phytoplankton?" *NASA Earth Observatory,* July 13, 2010, https://earthobservatory.nasa.gov/features/Phytoplankton.

17. John K. Brunson et al., "Biosynthesis of the Neurotoxin Domoic Acid in a Bloom-forming Diatom," *Science* 361, no. 6409 (2018): 1356–1358, doi: 10.1126 / science.aau0382.

18. Laurel Hamers, "This Hitchcock Movie Was Inspired by Crab Toxin Frenzy in Capitola," *Mercury News,* December 7, 2015, updated August 12, 2016, https://www.mercurynews.com/2015/12/07/this-hitchcock-movie-was-inspired-by-crab-toxin-frenzy-in-capitola/.

19. J. McQuaid, "Carbonate-sensitive Phytotransferrin Controls High-affinity Iron Uptake in Diatoms," *Nature* 555, no. 7697 (2018): 534–537, doi: 10.1038 / nature25982.

20. Lindsey and Scott, "What Are Phytoplankton?"

21. Lindsey and Scott, "What Are Phytoplankton?"

22. P. Tréguer et al., "Influence of Diatom Diversity on the Ocean Biological Carbon Pump," *Nature Geoscience* 11, no. 1 (2018): 27–37, doi: 10.1038 / s41561-017-0028-x; and

A. Mahadevan et al., "Eddy-driven Stratification Initiates North Atlantic Spring Phyto-plankton Blooms," *Science* 337, no. 6090 (2012): 54–58, doi: 10.1126 / science.1218740.

23.   Cavicchioli et al., "Scientists' Warning to Humanity."

24.   "The Effects: Dead Zones and Harmful Algal Blooms," US Environmental Protection Agency, January 31, 2022, https://www.epa.gov/nutrientpollution/effects-dead-zones-and-harmful-algal-blooms.

25.   "Large 'Deadzone' Measured in Gulf of Mexico," National Oceanic and Atmospheric Administration, August 1, 2019, https://www.noaa.gov/media-release/large-dead-zone-measured-in-gulf-of-mexico.

26.   Jeff Nesbit, "Oxygen Levels in Oceans Are Dropping Dangerously," *US News,* January 26, 2018.

27.   Sunke Schmidtko, Lothar Stramma, and Martin Visbeck, "Decline in Global Oceanic Oxygen Content during the Past Five Decades," *Nature* 542, no. 7641 (2017): 335–339, doi: 10.1038 / nature21399.

28.   Schmidtko et al., "Decline in Global Oceanic Oxygen Content."

29.   D. G. Bourne, K. M. Morrow, and N. S. Webster, "Insights into the Coral Microbiome: Underpinning the Health and Resilience of Reef Ecosystems," *Annual Review of Microbiology* 70, no. 1 (2016): 317–340, doi: 10.1146 / annurev-micro-102215-095440.

30.   O. Ladrière et al., "Morphological Alterations of Zooxanthellae in Bleached Cnidarian Hosts," *Cahiers de Biologie Marine* 49, no. 3 (2008): 215–227.

31.   "Coral Bleaching Threat Increasing in Western Atlantic and Pacific Oceans," NOAA Coral Reef Conservation Program, July 2015, https://coralreef.noaa.gov/aboutcrcp/news/featuredstories/jul15/bleaching.html.

32.   Katie Medlock, "Microplastics Are Killing Fish Faster Than They Can Reproduce," *Inhabitat,* June 8, 2016, https://inhabitat.com/microplastics-are-killing-fish-faster-than-they-can-reproduce/.

33.   Katsiaryna Pabortsava and Richard S. Lampitt, "High Concentrations of Plastic Hidden beneath the Surface of the Atlantic Ocean," *Nature Communications* 11, no. 1 (2020), 4073, doi: 10.1038 / s41467-020-17932-9.

34.   Oona M. Lönnstedt and Peter Eklöv, "Environmentally Relevant Concentrations of Microplastic Particles Influence Larval Fish Ecology," *Science* 352, no. 6290 (2016): 1213–1216, doi: 10.1126 / science.aad8828.

35.   Sinem Zeytin et al., "Quantifying Microplastic Translocation from Feed to the Fillet in European Sea Bass *Dicentrarchus labrax*," *Marine Pollution Bulletin* 156 (2020), 111210, doi: 10.1016/j.marpolbul.2020.111210; Hayley McIlwraith, "I Eat Fish, Am I Eating Microplastics?" Ocean Conservancy Blog, October 18, 2021, https://oceanconservancy.org/blog/2021/10/18/eating-microplastics/.

36.   "Overview of the Clean Air Act and Air Pollution," US Environmental Protection Agency, last updated August 10, 2022, https://www.epa.gov/clean-air-act-overview.

37.   "What's the State of Your Air?" American Lung Association, 2022, accessed October 15, 2022, https://www.lung.org/research/sota.

38.   "How's My Waterway?" US Environmental Protection Agency, Water Quality Assessment and TMDL Information, https://mywaterway.epa.gov/national.

39.   David Ewing Duncan, "Chemicals within Us," *National Geographic*, September 2006, 116–135.

40.   Intergovernmental Panel on Climate Change (IPCC), "Summary for Policy-makers," in *Global Warming of 1.5°C. An IPCC Special Report*, ed. Valerie Masson-Delmotte et al. (Cambridge: Cambridge University Press, 2022), DOI: https://doi.org/10.1017/9781009157940.001.

41.   H. O. Smith, J. D. Peterson, K. E. Nelson, et al., "DNA Sequence of Both Chromosomes of the Cholera Pathogen *Vibrio cholerae*," *Nature* 406, no. 6795 (2000): 477–483, doi: 10.1038/35020000.

42.   Bijal Trivedi, "Profile of Rino Rappuoli," *Proceedings of the National Academy of Sciences USA* 103, no. 29 (2006): 10831–10833, doi: 10.1073/pnas.0604892103.

Epilogue

Epigraph: Stephen Jay Gould, *The Richness of Life: The Essential Stephen Jay Gould* (New York: W. W. Norton, 2007), 214.

1.   "What Does It Mean to Be Human?" Smithsonian National Museum of Natural History, last updated August 15, 2022, https://humanorigins.si.edu/evidence/genetics.

2.   Shannon J. Williamson et al., "The *Sorcerer II* Global Ocean Sampling Expedition: Metagenomic Characterization of Viruses within Aquatic Microbial Samples," *PLoS One* 3, no. 1 (2008), e1456, doi: 10.1371/journal.pone.0001456.

Index

Page numbers in italic refer to photos and illustrations.